A REGION IN TRANSITION

To the memory of R.M. 'Meff' Hughes,
1910-1999, Durham miner

A Region in Transition
North East England at the millennium

Edited by

JOHN TOMANEY
Centre for Urban and Regional Development Studies
University of Newcastle, Newcastle upon Tyne
United Kingdom

NEIL WARD
Department of Geography, University of Newcastle
Newcastle upon Tyne, United Kingdom

LONDON AND NEW YORK

First published 2001 by Ashgate Publishing

Published 2017 by Routledge
2 Park Square, Milton Park, Abingdon, Oxon OX14 4RN
711 Third Avenue, New York, NY 10017, USA

First issued in paperback 2017

Routledge is an imprint of the Taylor & Francis Group, an informa business

British Library Cataloguing in Publication Data
A region in transition : North East England at the
    millennium. - (Urban and regional planning and development)
    1. Regional planning - England, North East 2. England, North
    East - Economic policy 3. England, North East - Politics and
    government
    I. Tomaney, John, 1963- II. Ward, Neil
    338.9'428

Library of Congress Control Number: 00 -134811

ISBN 13: 978-1-138-26370-3 (pbk)
ISBN 13: 978-0-7546-1022-9 (hbk)

Contents

PART III: ENVIRONMENT AND COUNTRYSIDE IN TRANSITION

List of Contributors

Susan Baines	Centre for Urban and Regional Development Studies, University of Newcastle
Paul Benneworth	Centre for Urban and Regional Development Studies, University of Newcastle
Priscilla Boniface	Formerly University of Northumbria
David Charles	Centre for Urban and Regional Development Studies, University of Newcastle
James Cornford	Centre for Urban and Regional Development Studies, University of Newcastle
Peter Fowler	Formerly University of Newcastle
Robert Hollands	Department of Social Policy, University of Newcastle
Chris Lanigan	District Audit/Audit Commission, Gateshead, formerly Department of Politics, University of Newcastle
Philip Lowe	Centre for Rural Economy, Department of Agricultural Economics & Food Marketing, University of Newcastle
Suzanne Moffatt	Department of Epidemiology and Public Health, University of Newcastle

Peter Phillimore   Department of Social Policy, University of
                             Newcastle

Andy Pike   Centre for Urban and Regional
                             Development Studies, University of
                             Newcastle

Tanja Pless-Mulloli   Department of Epidemiology and Public
                             Health, University of Newcastle

Mike Robinson   Centre for Travel and Tourism, University
                             of Northumbria

John Tomaney   Centre for Urban and Regional
                             Development Studies, University of
                             Newcastle

Neil Ward   Department of Geography, University of
                             Newcastle

Jane Wheelock   Department of Social Policy, University of
                             Newcastle

Rachel Woodward   Centre for Rural Economy, Department of
                             Agricultural Economics & Food
                             Marketing, University of Newcastle

Acknowledgements

We would like to thank the University of Newcastle's Institute for Urban and Rural Research for providing financial support for the research seminar at which the chapters in this volume were first presented and to all those who participated in the event. We would also like to express our gratitude to Sue Robson at the Centre for Urban and Regional Development Studies for her time and patience in assisting us with the editing and page-setting of the book.

John Tomaney and Neil Ward

1 Locating the Region: An Introduction

JOHN TOMANEY AND NEIL WARD

Introduction

This book is about is the North East of England. The production of the book is the result of an experiment conducted within a context of rapid economic, social, cultural and political change. It has its intellectual origins in its editors' interest in debates about the 're-emergence' of the region as basic unit of economic analysis and as the sphere most suited to the to the interaction of political, social and economic processes in an era of 'globalisation'. It is also borne out of an interest in the changing pattern of territorial relationships in the United Kingdom and the implications of this for the English regions in general and the North East of England in particular. While great claims are being made about the importance of regions, and new institutions are being created purportedly to serve their interests, there is a lamentably low level of debate within the English regions about the questions that arise from these processes. Lacking civic arenas in which the key issues can be raised and, more importantly, debated, the English regions are often poorly placed to grasp the scale of the challenges that confront them in this new era. However, questions are asked, information is gathered and, occasionally, solutions are proposed to regional problems among the universities. As a recent report on higher education noted, UK universities have had only very limited engagement with their wider regions (Dearing Committee, 1997, para 12.7). Although English universities, especially in regions like the North East, are frequently undertaking empirical research within their regions — sometimes for regional agencies — there appear to be few attempts to bring this research together and to assess their combined contribution to understanding development.

The papers in this volume, therefore, result from an effort to redress this situation. Their origins lie in a seminar held at Newcastle University in 1998, designed to assess (and demonstrate) the breadth of research with a regional focus. The meeting involved researchers from six of the University's nine faculties, although the focus was strongly a social science one. One aim of the meeting was to bring together diverse researchers whose work was focused on the North East region or some geographical component of it. A second hope was that common themes and insights would emerge from the discussions. A third aim was to make at least a modest contribution to the University's objective of strengthening and focussing its engagement with wider interests in the North East beyond the ivory towers of academia.

The remainder of this introductory chapter provides an introduction to both the claims about the importance of 'the region' and regionalism, especially the implications of this in an English context, and to the chapters arising from the seminar.

Locating the region

There is a heightened interest in 'the region' as a locus of political, economic and cultural activity. A large amount of literature suggests that, for a variety of reasons, 'the regional question' has gained a new prescience. In the UK, an obvious source of interest derives from recent constitutional changes that have led to the creation of devolved political institutions in Scotland, Wales, Northern Ireland. But for many scholars regionalisation is a 'global' or, at least, 'Europe-wide' process. From this latter perspective, recent events in the UK simply reflect similar events occurring elsewhere in the world. Regionalism, therefore, is seen as one aspect of a broad set of economic, social, cultural and political changes that are transforming territorial relationships. In particular, according to this position, a defining characteristic of the contemporary period is the declining capacities of the nation-state. It is claimed that the power and authority of the nation-state has been eroded from a number of directions. First, it has been challenged from above by 'globalisation', including the globalisation of economic activity and the growth of supra-national political forms of authority such as the EU, which regulate that activity. But the state is also under challenge laterally by the further advance of market relationships. The resurgence of the private sector and the alleged rise of the 'knowledge economy' are linked to, and encourage, economic globalisation. Taken together 'globalisation' and 'marketisation' of social and economic relationships erode the capacities of the state in economic management, in social solidarity, in culture and identity formation, as well as its institutional configuration (e.g. Giddens, 1998; Castells, 1996; Held, *et al.*, 1999).

In addition, there is evidence that nation-states are being challenged from below by the assertion of regional interests. These regional claims can be seen as both a component of the broader challenges faced by the state and, at the same time, as a response to them. Whereas throughout most of the post-war period governments intervened in the economy to promote balanced forms of regional development, economic modernisation and territorial integration, latterly these types of actions have been less of a priority for the state. This leaves regions more directly exposed to international forms of competition, while states no longer monopolise the external links of regions. Regions become actors in their own right. So, according to one proponent of this perspective:

> This has produced a new regionalism marked by two linked features: it is not contained within the framework of the nation-state; and it pits regions against each other in a competitive mode, rather than providing complementary roles for them in a national division of labour (Keating, 1998, p. 73).

The suggestion that there is a 'new regionalism' is now commonly made and widely accepted in academic, media and political circles and beyond (e.g. Harvie, 1994). On the face of it, the evidence for this is compelling and can be identified in the economic, cultural, environmental and political fields of action.

'New regionalism' and the economy

The current phase of economic restructuring is typically seen as a significant factor promoting a new role for regions (e.g. Amin, 1994). The relationship between economic change and territorial development has moved closer to centre stage in social science. This represents an important change in that the main bodies of economic thought hitherto have shown little concern with the regional question.

Neo-classical economics historically showed little concern with regional problems. It sees the market as self-equilibrating and, as a result, regional problems are no more than the spatial manifestation of an adjustment failure on the part of the factors of production. This viewpoint was certainly reflected in the regional economic policies of some governments in the 1980s, notably the Thatcher governments in the UK. More recently, equilibrium-based interpretations of economic development have come under attack again mainly as a result of the currency instabilities of the late 1990s. In this context, Keynesian analysts have mounted a counter-offensive. The Keynesian approach conceived of economic activity as a disequilibrium process and identified a key role for governments in managing the tendency toward imbalance. Surprisingly, given the regional concentration of unemployment in the inter-war period, Keynes did not write about regional problems as such, nor did his policy recommendations contain any suggestions on regional policy. In the post-war period, however, essential Keynesian insights were used to inform not only general approaches to macroeconomic management on the part of governments, but also the management of regional problems. One the one hand, it was accepted that: '[...] the play of forces in the market normally tends to increase, rather than to decrease, the inequalities between regions' (Myrdal, 1957, p.26). On the other hand, the regional problem tended to be conceived in typically Keynesian terms as a demand problem:

> ... why does the rate of growth of labour, capital and total factor productivity differ between regions? The major explanation must lie in differences in the strength of demand for regions' products. The only true supply constraint on growth is land-based resources, but economic activity in most regions in mature economies is not land-based (Thirlwall, 1980, p.419).

This 'spatial Keynesianism' (Martin, 1993) marked the post-war approach to regional policy that involved the improvement of infrastructure in assisted regions and the provision of incentives to mobile investors or government support to existing firms. The aim of these policies, together with restriants on growth in core regions, was to redistribute investment to regions with deficient demand. These policies were generally successful in creating jobs outside the core regions of states (see Moore *et al.*, 1986, for evidence from the UK) and were addressed to the modernisation of both rural and declining industrial regions. However, they were criticised for creating branch plant economies in some regions and failing to tackle underlying productivity, technology and entrepreneurship problems.

Such policies, though, have declined in importance in the last quarter of the twentieth century. Advocates of the 'new regionalism' relate this decline in part to the constraints on the power of the nation-state. States, they suggest, no longer have the capacity or resources to support large-scale regional policies because their policies must first pay attention to the increased power of the international financial system or transnational companies rather than domestic constituencies. However, the new regionalism also responds to a new logic of economic restructuring. For some writers, this new logic derives from changes in macro-economic structures that have led to the emergence of new modes of growth, such as the transition from Fordism (standardised, mass production) to post-

Fordism (customised, small batch production), or the rise of a knowledge or information-based society (Amin, 1994; Best, 1990; Castells, 1996).

New forms of industrial organisation have proved themselves adept at coping with these changed market conditions. For instance, during the 1980s and 1990s much attention was visited on the industrial districts of the Third Italy, a set of regions that saw a sustained improvement in growth relative to Europe as a whole. Numerous analyses have shown that the competitive advantage of these districts derives from the formation of networks of firms with a highly specialised division of labour, typically in traditional industrial sectors. A degree of firm specialisation means that the production process is contained within inter-firm linkages that guarantee external and scope economies, often using new technologies. The nature of the inter-firm division of labour generates powerful agglomeration effects, hence the use of the term 'industrial district'. Firms both compete and collaborate with each other – a situation that comes about because of the mutual trust of a shared craft and industry culture. Subsequent analyses have identified a range of 'new production spaces' that, while differing in detail from the Italian example, appear to have features in common. In short, there appears to be something about locally constituted production systems that generates competitive advantages.

Storper (1998) has referred to the importance of 'untraded interdependencies' that operate at the regional level, while Cooke and Morgan (1998) have identified an 'associational economy' or 'network economy' rooted in regional production systems. Regional institutions often act to underpin these new production spaces by facilitating networks that promote upgrading of supply-side conditions. These policies are frequently presented as offering an alternative to (or, perhaps, a 'Third Way' between) free market and Keynesian approaches to regional development (Amin, 1999; see also Martin, forthcoming). Hence the literature notes that strong regional government often is associated with these new production spaces (Cooke and Morgan, 1994). There exists barely a government in the industrial world that has not at least paid lip service to these ideas in the development of their regional policies. In addition, supra-national organisations such as the EU, OECD and World Bank have been strong supporters of versions of the same thesis (*e.g* Javed Burki, *et al.*, 1999).

However, there are good reasons to doubt aspects of this proposition. While few would deny that significant economic changes occurred in the last quarter of the 20th century, many scholars have tended to overlook the existence of strong continuities between 'old' and 'new' forms of industrial organisation, especially as far as their economic geography is concerned. Moreover, even where phenomena such as the 'new industrial spaces' can be identified it is far from clear whether they signal the emergence of a 'new paradigm'. The barriers to emulation of these systems are high, not least because they probably owe a large part of their dynamism to institutional and cultural structures that have been built up over a long period. Moreover, even among the new industrial spaces there are many differences, often reflecting their particular national setting, making it difficult to derive a general account of their characteristics. To date there have been few efforts to undertake a general analysis of the dynamics of regional growth taking into account economic, social and political factors. Even now it is not clear whether they represent a contingent response to turbulent economic conditions in the late 20th century or a new model of economic development for the 21st. Certainly, there is evidence that large firms, far from being eclipsed by networks of small firms, are reasserting their power and, as a result, retain their ability to shape regional fortunes. There is also compelling evidence pointing to the enduring importance of the national context in shaping the terms of regional development (see Harrison, 1997, Lovering, 1999; Rodríguez-Pose, 1998). In short, while there have been significant economic changes it is very difficult to generalise about their impacts on regions.

'New regionalism' and the environment

A further factor stimulating interest in the importance of the region has been the growing interest in environmental sustainability. The notion that our planet faces an environmental crisis is now widely held, even if the meaning of this crisis and the appropriate responses to this are fiercely debated. Certainly, the growth of industry and associated patterns of consumption in urbanised, developed societies presents a challenge for sustainable development. At one level, the major issues of the effects of the production of energy from fossil fuels, population growth and large-scale deforestation are inter-related, *global* problems. Yet at the same time, the solutions to ecological pressures are often identified as being located at regional scale (see, for example, Douthwaite, 1996).

For instance, one of the most controversial issues for high-consumption developed countries concerns the management of wastes and effluents. Incineration and landfill have been the preferred solutions to the problem of waste in the past. The case against incineration though, is that it causes pollution (especially because of the toxicity of modern waste) and resource depletion. There is a strong case, on the other hand, for recycling of a much greater proportion of waste material than is currently the case.

Moreover, there is compelling evidence that the regional scale is the one most appropriate for this to take place. Municipal and regional authorities have been the leaders in new waste management strategies in the United States and the EU.

The environmentalists' famous slogan is 'think global, act local'. Yet, the space for local environmental initiatives is constrained by larger national and international forces. The environment is big business and large corporations have proved adept at shaping agendas to their needs. Where ecological problems are greatest, even in developed societies, there is typically limited access to the capital needed for the kinds of alternative strategies that may hold long-term environmental and economic benefits. Finally, public policy, and especially the regulatory activities of national governments and international bodies such as the EU, continues to be critical to defining the space for local action on the environment.

It may be, however, that it is the unintended political consequences of rising environmental consciousness that have the biggest impact on the regional arena. The shift to the politics of environmental sustainability appears to promote regional consciousness:

> In the world of industry, what is most willingly condemned is the proliferation of pollution and the often serious damage done to the environment as a result. In response there are demands for a rehabilitation of nature. Immediately spatial groups assume a new significance: one must unite to save the rivers and streams, limit the emission of gasses causing harmful acid rain, avoid the concreting over of the shores and the harmful consequences of mass tourism. The new regional consciousness colours itself green (Claval, 1998, p. 155).

Culture, identity and the 'new regionalism'

The bases of regional consciousness are worth further consideration here. Questions of regional identity are central to claims about the new regionalism. They already appear in the discussions about any new paradigm of regional economic development and the growth of environmental concerns described above. For instance, shared craft and industrial values are seen as being critical to the promotion of the trust-based forms of inter-firm relationships characteristic of places like the Third Italy. Indeed, reviewing the experience of the Third Italy and similar regions, Cooke and Morgan have argued that "...contemporary regional economic success is inseparable from cultural, social and institutional accomplishment" (1994, p. 91).

The assertion of regional identity frequently is seen as a recent phenomenon, reflecting the growing salience of the region as an economic and political actor. In the past, the growth of nation-states, especially in the era of the welfare state, was often accompanied by strong forces for uniformity in the cultural arena. However, according to one perspective, the erosion of national cultures in part results from the current pressures on the nation-state that arise from economic changes and the emergence of supra-national forms of political authority. The spread of mass communications, the convergence of information and communications technology and the associated growth of the global media industries have all contributed to the expansion of global cultural processes. Giddens specifically links the rise of regionalism to the emergence of globalization:

> Globalization 'pulls away' from the nation-state in the sense that some powers nations used to possess, including those that underlay Keynesian economic management, have been weakened. However, globalization also 'pushes down' – it creates new demands and new possibilities for regenerating local identities. The recent upsurge of Scottish nationalism in the UK shouldn't be seen as an isolated example. It is a response to the same structural processes at work elsewhere, such as those in Quebec or Catalonia (1998, pp. 31-2).

Regional identity is difficult to define and measure, but at an anecdotal level the evidence for an explosion of interest in cultural regionalism is available in Europe (Bassand, 1993). Regions in this perspective are social constructions and reflect the search for new forms of identification in the face of the erosion of national affinities (Claval, 1998). In the social construction of new regional identities the uses of history are important. Interpretation of the past is a key element of cultural representation. This can have a romanticised aspect as can be evidenced in Britain by the growth of the heritage industry. However, as Keating notes:

> There is also, however, a search for a 'usable past', a set of historical referents which can guide a regional society on its distinct road to modernization, bridging the past, via the present, with the future. The revival of serious regional history and the challenging of dominant national interpretations then become an instrument in guiding a regional society to its own future (1998, p. 84).

In many parts of Europe regional identities are supported by tendencies that have strengthened regional print and broadcast media systems. Federal states such as Germany or Switzerland, for instance, have had regional television systems for some time. States that have moved in the direction of political regionalism often support it by restructuring their broadcasting systems along similar lines. For instance, once the autonomy of Catalonia was recognised by the Spanish state, the autonomous community established a new television channel broadcasting in the Catalan language. According to Bassand:

> Notwithstanding the very high cost of the operation, people in Catalonia agreed that it was worthwhile as an affirmation of Catalan identity, which is rooted in the Catalan language. The large numbers of migrants living in this region are being 'Catalanised' by television, and social contact in Catalan between the other inhabitants is facilitated (1993, p.137).

Similar processes have accompanied — and encouraged — the growing political autonomy of Wales and Scotland in the UK.

Although many claims are made about the resurgence of regional identity as a general process affecting all societies in the face of globalisation, the issue remains difficult to grasp analytically. It is difficult to come up with a working definition of regional identity, especially one that lends itself to comparative analysis. An additional problem lies in assessing how such identities affect collective behaviour and political action. It is by no means certain that a shared regional identity should have political consequences. Neither is it inevitable that regional identities should replace national ones as the principle mode of cultural and political mobilisation. National identities remain extremely potent in politics, as debates about further European integration reveal. Moreover, where regional identity does take a political dimension, its form can be unpredictable. For instance, in France the strong identity of the Alsace region paradoxically has produced not a regionalist movement, but for a variety of reasons a high level of support for the Front National (Bihir, 1998).

Notwithstanding these caveats, a complex amalgam of factors is ensuring that, in some contexts, regional identities are gaining salience. However, it is equally important to recognise the enduring importance of the national context in framing the conditions for the emergence of regionalism.

'New regionalism' and politics

> The development of feelings of belonging to a region, in old developed countries like France, appeared initially as a personal matter, a kind of fashion leading to a series of private initiatives in the domain of the environment and culture. But, very quickly states and public authorities became involved in these localized evolutions (Claval, 1998, pp. 283-4).

The existence of an elected regional tier of government is now a feature of every large Member State of the European Union. Even the UK, hitherto the exception to this rule, created a Scottish Parliament and Welsh Assembly in 1999. Although, the idea of a 'Europe of the Regions' is a relatively new one, regional government has existed in some form in several states for some time. Germany and Austria became federal states in the 1940s. Italy moved in the direction of regional government in the 1970s. However, the pace of regionalisation did appear to quicken during the 1980s with states such as Spain, France and Belgium creating regional tiers of representation.

One of the most commonly offered explanations for the recent growth of regional government stresses the role of economic restructuring in impelling the decentralisation of administrative and political power (Keating, 1998). This literature stresses the supportive role of regional governments in underpinning the performance of the 'new economic spaces' described earlier (*e.g.* Cooke and Morgan, 1998). Hausner describes the logic of this approach:

> an industrial policy oriented towards structural changes in the economy and the promotion of producers' adaptability to the conditions of domestic and international competition must focus on meso-level structures in the given economy and its social environment. Therefore, the creation of intermediate level structures that would facilitate economic restructuring is the top priority of industrial policy and the goal of the economic strategy (1995, p. 261).

General claims about the role of decentralised institutions in the promotion of regional growth need to be treated with caution, however. Many of the claims about the effectiveness of local institutions are drawn from a narrow range of case studies. Indeed, more systematic attempts to assess the impact of regional institutions on growth on a Europe-wide basis suggest a less clear picture, where the determinants of growth remain capital and technology that, with some exceptions, remain concentrated in the same regions as before. Thus, 'new' patterns of regional growth, in the main, are rooted in older ones (Rodríguez-Pose, 1998).

However, not all accounts of the new regionalism emphasise the centrality of economic restructuring in their explanations. Other accounts emphasise a more general problem of democratic deficits that occur as a result of the diminished power of the nation-state in the face of globalisation. Globalisation is accompanied by a general disenchantment with normal channels of politics. Specifically, a gap has emerged between the system of representation through state institutions, and decision-making that has retreated into technical and social networks. This leads to a divorce between 'politics' and public policy, which by implication can be filled by regional democracy. The advance of regional democracy then reflects a more widespread effort to reinvigorate democratic politics and civil society (Giddens, 1998).

> Despite the political appeal to some of a 'Europe of the Regions', it is not entirely clear how well founded it is as a concept. Keating (1998) has recently produced an overview of the emergence of the 'new regionalism in Western Europe'. Coverage, though, is uneven and this reveals a larger problem. We learn about Belgium, for instance, but not about its larger neighbour, the Netherlands, where regionalism is less important. Similarly, we read about Spain but not about Portugal, where the population has rejected regionalisation. Within the UK, we discover much about the position of Scotland but, curiously, much less about Northern Ireland. The examples of the new regionalism are drawn from the most favourable national contexts, although even here the meaning of regionalism varies widely. This raises the question about what is 'new' and what is 'European' about the new European regionalism.

Reviewing the terrain of the new regionalism one is struck, despite the evidence of regional assertion, by the enduring role of the national context in shaping patterns of regionalism. A recent review of regional growth performance in Europe concluded:

> Globalization of the world economy, flexibilization of the production system, the current processes of political integration and deregulation trends might have contributed to bring regions and the regional dimensions to the fore. On a cross-regional level, however, this phenomenon is far from being widespread. Regional growth patterns in Europe during the 1980s are still as bound by the national dimension as they were during the zenith of the mass production era in the 1960s. The greatest diversity in social and political data is found across the nations and not across regions within a nation. Hence, it could be stated that economic, social and political indicators are greatly determined by the national dimension, and the nation-state still constitutes a very solid unit for the analysis of current social, political and economic transformations. Regional specific characteristics and diversity only come to light when social and political variables are nationally weighted (Rodríguez-Pose, 1998, pp. 230-231).

All this is not to deny that a process of regionalisation may be occurring, or that the process is unimportant. Rather it is to emphasise the uneven and path-dependant nature of change – in which the national dimension remains of central explanatory importance — and to caution against viewing regionalisation as the outcome of an unfolding universal logic of territorial restructuring. An examination of the uneven pattern of English regionalism amply demonstrates the value of such a perspective.

In what sense an English region?

> [...] we are not a regional nation. In fact, if you put the different TV companies, police authorities, government offices, water companies and electricity provides in a room together, none of them would be able be able to agree on where our regions are supposed to be.
> (Rt. Hon William Hague, MP, 'Identity and the British Way', speech to the Centre for Policy Studies, London, 19th January 1999).

England is an anomaly in relation to the 'new regionalism'. It is an axiom of the debate we have described that England remains unfriendly toward the charms of the new regionalism. Harvie (1991), for example, described English regionalism as 'the dog that never barked'. The contributions to this book in various ways take issue with the view of English regionalism as a non-issue. The focus of the chapters is on the distinctive aspects of the economy, environment and politics of the North East region. It would be surprising indeed to find that a trend as allegedly generalised, as the 'new regionalism' had made no impact on England. Yet, as these chapters show, regional development in its economic, social, cultural and political forms is complex and problematic and cannot be reduced to the unfolding logic of the 'new regionalism'.

England contains distinctive economic regions, but this distinction owes little to their contrasting patterns of 'un-traded interdependencies', as suggested by the new regionalists cited above, but rather more to underlying economic inequalities. These are revealed by making a broad comparison of the socio-economic structure of the North East and the South East. The South East of England contains the headquarters of banks and large firms, a disproportionate share of research and development activities and fast growing business service and hi-tech enterprises. Its social structure is biased toward higher income and professional groups. By contrast, the North East, in social and economic terms, is characterised by bias toward manufacturing (which is largely externally controlled) and under-developed business service and hi-tech sectors. It has low levels of research and development and a poor record of new firm formation. Its social structure is biased toward lower income and non-professional groups. Above all it is characterised by chronically high levels of unemployment, with especially severe problems of long-term and 'hidden' forms of unemployment. The region's contemporary social and economic weaknesses reflect a long pattern of industrialisation and de-industrialisation that, among other things, have left a legacy of environmental and associated health problems. The region's economic performance is heavily conditioned by the decisions of externally owned firms and macro-economic processes over which it has little control. By now, the North East is caught is a vicious circle of decline and faces increasing difficulties in closing the prosperity gap with the South. This picture of regional disparity is painted in broad brush-strokes — for instance, it ignores intra-regional inequalities that are present in different ways in both regions — but it can be generally supported by official statistics. Ironically, it is these conditions — and the sense of economic injustice they generate — which underpin the region's recent assertion of its cultural and political identity.

The discussion of identity within England rarely accords much attention to the question of regions. Simultaneously, though, the dominant images of 'Englishness' refer mainly to the south of England. In his recent book on *The English*, Jeremy Paxman (1998) expresses the issue appositely:

> If you had to guess the whereabouts of the lane, small cottage and field of grain, where there'll always be an England, you'd decide pretty quickly where it was *not*. You could instantly rule out places like Northumberland and Yorkshire, where fields would have dry-stone walls and are more likely to be full of sheep anyway (1998, p. 156).

Paxman follows a familiar pattern in discussions of English identity insofar as the 'north' is an aspect of the 'other'. Indeed, Peter Taylor (1993) has referred to the North as 'England's foreign country within'.

But the English regions can be argued to have their own distinctive identities. In the case of the North East, the region's modern culture owes a great deal to the pattern of industrialisation and de-industrialisation in the 19th and 20th centuries. One consequence of this economic history was to lay the foundations for the emergence of a regionally distinctive working class culture and politics based around 'Labourism'. However, the region can be said to have a distinctive identity throughout its history, deriving both from its position as the core of the kingdom of Northumbria in the ninth and tenth centuries and, in the Middle Ages, as the 'debatable lands' between Scotland and England. This aspect of the region's development means that, for instance, the region's rural areas have always born more resemblance to neighbouring areas of Scotland, than to England's green and pleasant lands in the South. Colley, in her study of 18th century Britain, observes;

> Northumberland, for instance in the way that its people looked and lived and thought, was much closer to being a Scottish than an English county. Here, as in the Scottish lowlands, the poor consumed oatmeal as a matter of course, a cereal that — as Samuel Johnson remarked in his famous dictionary — more affluent southerners dismissed as animal fodder. Here, too, over a third of all adults may have been able to read by the early 1700s. This was virtually the same level of literacy as existed in Lowland Scotland, but at a much higher level than, say, in the English Midlands where Johnson hailed from. Books and newspapers from Scottish presses were far more common in Northumberland than London-produced reading matter, and Scots and their accents were infinitely more familiar than visitors from the South. Northumbrians and Lowland Scots even tended to look alike, with the same raw, high-boned faces and the same, angular physiques. 'To pass from the borders of Scotland into Northumberland', a Scottish clergyman would write at the end of the 18th century, was rather p. 16).

The contemporary assertion of the region's identity reflects a complex amalgam of a 'reinvented' regional history (including, for instance, a campaign to 'return' the ninth century Lindisfarne gospels to the region from the British Museum where they now reside) and strong claims of economic injustice. These have intermingled with Labourist political traditions, and awareness of the impact of devolution on Scotland and Wales, to underpin claims for new regional political institutions (see Tomaney, 1999a). This dynamic of English regionalism has already had an impact on UK politics and is likely to do so into the future.

In general, however, English political culture provides unpropitious conditions for regionalism. Politics remains heavily focussed on the Westminster parliament, which has proved itself reluctant to cede power to the regions. The Labour Government in 1999 created Regional Development Agencies in each of the English regions. Ostensibly these were concerned with addressing the economic imbalances described above. However, they also reflected an attempt to pre-empt the rise of political regionalism in England as a response to devolution in Scotland and Wales (Tomaney 1999b). The Labour Government's proposals for England fall far short of those proposed for Scotland and Wales, and it remains to be seen whether they will accommodate the nascent English regionalism, or stimulate further demands for regional autonomy.

The evolution of regionalism in the North East and other parts of England will be determined in part by future political and economic processes in the EU and UK. However, patterns and processes of economic, social and cultural change within the English regions themselves will in turn affect the prospects for political regionalism in England. To coin a phrase, English regions will make their own histories, but not in circumstances of their own choosing. In making this point we are cautioning against adopting a simplistic view of the 'new regionalism' that reduces regional change to the unfolding of some global logic or new development model.

England then is not well served by explanations that define it as some distinctive exception to the new regionalism. In drawing attention to the path-dependent nature of regional change and the continuing importance of the national scene, we prefer to see England as simply another case, that needs to be explained in the context of its own historical and contingent circumstances. In short, there is a powerful case for further research into the social and political constitution of English regional life.

Within this context, academics have a huge opportunity — and even a social responsibility — to shed light on the evolving scene. This is particularly because in the English regional context, unlike say other EU Member States or even other parts of the UK, there is an absence of deeply rooted civic institutions within which these debates can be had, reflecting the centralisation of English political culture. It is partly in this spirit that this volume has been compiled.

Structure of the book

Following this Introduction, the book is divided into three main parts. First, the notion of an economic transition in the North East is critically explored. In Chapter 2, David Charles and Paul Benneworth examine the nature of economic change in the region since 1975, with a particular emphasis on the regional economy's links with other places. They reflect on the key question of whether the changes to the region's economic structure are indicative of some fundamental change in its position within national and international spatial divisions of labour, or whether they are simply a response to the changing scale of economic regulation. The chapter also includes a comparative analysis of the region's economic performance in the light of other European regions. It concludes that any future development strategy for the North East should seek both to attract high-order business functions to the region while, at the same time, fostering a more endogenous form of economic growth 'from within'.

In Chapter 3, John Tomaney, Andy Pike and James Cornford move on from this overview of the regional economy to examine an individual case study of industrial change. The case illustrates some of the implications of economic restructuring, not only for the functioning for labour markets in the region, but also for the lived experience of economic life for those workers made redundant from traditional industries. The chapter presents the findings from a survey of over 1,600 people made redundant from the Swan Hunter's shipyard between 1993 and 1995 to elicit their subsequent fortunes in the labour market. The survey revealed that under 40 per cent of the redundant workers had remained unemployed two years later, although only 10 per cent felt that their personal situation had improved since the closure, with 70 per cent feeling quite the opposite.

The rhetoric of economic transformation often emphasises notions of 'entrepreneurial culture', part of which involves the setting up of small firms. Indeed, a proportion of the Swan Hunter workers went on to become self-employed in some way.   However, research into the declining economic performance of industrial regions has often tended to dwell on the role of large manufacturing firms and the SME sector. It is only relatively recently that academic attention has begun to turn to the role of those smallest economic units — micro-businesses (defined as firms employing fewer than ten people), which are now estimated to account for around a third of employment in the EU. In their chapter, Sue Baines and Jane Wheelock present results of an empirical study of micro-business livelihoods. Their comparative analysis of firms in the North East and South East of England focuses, in particular, on the relationships between families and firms and the nature of recruitment. Their research suggests that, while politicians continually stress notions of entrepreneurship, dynamism and innovation in the small firm sector, empirical evidence suggests that small family firms are much more likely to characterise very old ways of working, not least in terms of gender relations. Notably, the relationships between 'family' and 'business' seem from the study to be broadly similar in the North East and the South East.

The second part of the volume turns from issues of economic change to a set of associated questions concerning political and cultural change. Studies of regional issues in England have, we would argue, tended to be dominated by concerns about economic structure and performance. Less attention has been focused upon the social and cultural aspects of the regional question, or on the *links between* economy, politics and culture in regions. In Chapter 5, Chris Lanigan draws on the research he conducted at Newcastle University's Department of Politics on regional politics and regional identity in the North East. He identifies two attempts to define particular conceptions of regional identity in the region — a market oriented identity project, and a regional community oriented identity project. The former is a type of economic 'boosterism' which emphasises the shift from a mining and manufacturing economy to one where services and, particularly, leisure industries are the new driving forces. The latter is rooted in the notion of a distinctive cultural identity in the North East. The chapter concludes by reflecting on the prospects of regional politics and suggests that differences between the North East and the rest of England are likely to become more strongly emphasised over time. Political devolution in Wales and, more particularly, just across the border in Scotland, will serve to encourage regional political actors in the North East to utilise regional identity and draw attention to the region's distinctiveness.

The analysis by Peter Fowler and colleagues in Chapter 6 parallels that by Chris Lanigan. They examine the nature of the relationship between economic change and regional culture. Here too, the authors develop a dualism of 'two cultures' as an heuristic device to understand the struggle to give meaning to economic change in the region. For them, the two cultures are the (old) regional culture and the (new) external culture. Interestingly, they show how the development of the latter, based on the globalising service sector and standardised cultures of consumption, still reflects crucial traits of the old North East. These commonalities include notions of economic dependency and a sense of external control over events. In the North East, the seeming paradox is that the region's people seem to prefer a culture that meshes better with the industrial past, yet they risk becoming passive observers as economic forces shape the region into a form 'like everywhere else'. As a programme of action, the authors of Chapter 6 call for an active engagement in the transition of economy and culture, and suggest that it is through resistence to the homogenising influences of commoditised consumption that a new and distinctive cultural economy of the North East might be forged.

In Chapter 7, Robert Hollands completes this section of the book on politics, economy and culture with a detailed examination of one sociological grouping that has been identified as a harbinger of the North East's cultural transition — young adults. For the practice of 'going out' and, more particularly, the cultural flamboyance of Newcastle's Bigg Market are frequently pointed to as symptomatic of the transformation of the region. His chapter provides a wide-ranging assessment of how economic change has influenced the relationships between young adults and the spheres of work, home and community as well as consumption patterns. It goes on to reflect on the implications of these changes for gender relations, regional identity, and the shifting political economy, of the transition from youth to adulthood.

The third and final part of the book turns to two aspects of social and economic life in the North East that, over recent decades, have attracted comparatively little attention from social science researchers — those of environmental quality and countryside management. This relative neglect is perhaps unsurprising for two reasons. First, pollution and other environmental problems have tended to be seen as the exclusive preserve of natural scientists who measure, monitor and model environmental processes and attempt to 'engineer' their solutions. Second, from a socio-economic perspective, it has been the profound changes in the structure and fortunes of key aspects of the region's economy that have given most cause for concern amongst the populace, politicians and policy-makers in the region. Of course, these concerns have a particular geographical expression, being concentrated in the urban and mining parts of the region. Understandably, the research agendas of social scientists will at least in part reflect these concerns and priorities.

However, in Chapter 8 Peter Phillimore and colleagues provide a good example of the contribution that social scientific analysis can bring to our understanding of the nature of environmental problems. Its focus is the issue of air pollution and air quality monitoring in Teesside, which may at first glance appear a simple technical matter. Yet, the authors show how politics and power interplay with the seemingly 'scientific' issue of air quality monitoring. They trace the evolution of such monitoring in Teesside and show how changes to the system have served to downplay the contribution of industrial emissions to air pollution — a source of some difficulty for the image Teesside presents to the outside world — and increase the profile of road traffic emissions. Changes in monitoring, and the interpretation placed upon monitoring results, gives the impression that, in terms of air quality, Teesside is much like other urban centres in Britain. The chapter goes on to highlight the efforts of local authorities and industries, through leafleting Teesside households, to promote the idea that "wrong perceptions" about poor air quality in Teesside abound, and these pose a problem for the area's wider reputation and threaten investment. The leaflets seek to encourage people to "get together and challenge wrong perceptions".

It is clear from the chapter that what counts as acceptable local environmental quality is contestable between different social interests, and the way knowledge about environmental quality is produced is often part of that struggle. It is only through careful scrutiny of how environmental information is produced and made use of, that we can adequately make sense of those claims about environmental 'improvements', which often accompany industrial restructuring and technological change.

In Chapter 9, Neil Ward and Philip Lowe look to the North East's countryside and explore the fortunes and contribution of rural areas within the regional development agenda. The chapter argues that regional development research and policy debates have tended to be primarily concerned with 'cities and their regions'. However, the establishment of the new Regional Development Agencies (RDAs), coupled with European reforms to the Structural Funds and Common Agricultural Policy, are serving to reposition rural areas within regional development agendas. The new RDAs have an explicit remit for the economic development of their regions as a whole, and are required to draw up rural development programmes. In addition, the direction of Structural Fund and CAP reform is towards a more integrated and territorial approach to rural development in regions and a move away from the dominance of a highly sectoral agricultural policy. The socio-economic structure of the rural parts of the North East is distinctive in English terms but, the chapter concludes, a 'regional rural policy' should build on the inherent competitive strengths, be they in tourism, regionally-branded foods, or new approaches to agricultural development.

In Chapter 10, Rachel Woodward analyses a case study of a particular and contentious development question in the North East's countryside — that of the role of military training in Northumberland National Park. The army has been training at the Otterburn Training Area since the time of the First World War, and in 1995 applied for planning permission to develop the site and intensify its usage for training with heavy artillery weapons systems. The subsequent planning saga brought to light a whole host of arguments about the relative costs and benefits of such training and its appropriateness in a National Park. In the chapter, three particular arguments about land use are examined. The first surrounds the local economic significance of a large military installation in a relatively isolated rural area. The second concerns the regional significance of the site from a landscape and conservation perspective. The third centres on the national significance attributed to the area by the Ministry of Defence.

The case study helps to show how the identity and characteristics of places are constructed and struggled over by different social interests, and highlights how the meanings we give to places are never immutable. Of course, this lesson applies as much to the North East region as a whole as it does to the Otterburn Training Area. Indeed, part of the task of this volume is to contribute to a debate about the contemporary fortunes of the North East, and about change and continuity in those fortunes, and to shed light on how new and old meanings are forged, maintained or contested.

Different aspects of these conflicts can be read through the chapters that follow and prompt a range of questions. Will the North East remain cast in its role as a declining industrial region or recreate itself as a pioneer of the new service economy? Does the demand for greater political autonomy represent an attempt to advance a new political identity or retreat into an old one? How might economic or environmental changes alter the relationships between the urban, the rural and the regional? Who might be the winners and losers from the material outcomes of these struggles? It is hoped that the contributions that follow shed at least some light on these questions.

References

Amin, A. (1994) *Post-Fordism*. Oxford: Blackwell.

Amin, A. (1999) 'An institutionalist perspective on regional economic development', *International Journal of Urban and Regional Research*, 23: 365-378.

Bassand, M. (1993) *Culture and Regions of Europe*. Strasbourg: Council of Europe Press.

Best, M. (1990) *The New Competition*. Cambridge: Polity Press.

Bihir, A. (1998) 'Exceptionne alsacienne', *Le Monde Diplomatique*, May, pp.16-17.

Castells, M. (1996) *The Rise of the Network Society*, Oxford: Blackwell.

Clavel (1998) *An Introduction to Regional Geography*. Oxford: Blackwell.

Colley, L. (1992) *Britons: Forging the Nation 1707-1837*, London: Pimlico.

Cooke, P. and Morgan, K. (1994) 'Growth regions under duress: renewal strategies in Baden Württemburg and Emilia Romagna', In Amin, A Thrift, N (Eds.) *Globalization, Institutions, and Regional Development in Europe*. Oxford: Oxford University Press.

Cooke, P. and Morgan, K. (1998) *The Associational Economy*. Oxford: Oxford University Press.

Dearing Committee (National Committee of Inquiry into Higher Education) (1997) *Higher Education in the Learning Society*, London: Stationary Office.

Douthwaite, R. (1996) *Short Circuit: Strengthening Local Economies for Security in an Unstable World*. Dublin: Lilliput.

Giddens, A. (1998) *The Third Way*. Cambridge: Polity Press.

Harrison, B. (1997) *Lean and Mean*. New York: Guilford.

Harvie, C (1994) *The Rise of Regional Europe*. London: Routledge.

Hausner, J. (1995) 'Imperative versus interactive strategy of systematic change in central and eastern Europe', *Review of International Political Economy* 2, 249-66.

Held, D., McGrew, A., Goldblatt, D. and Perraton, J. (1999) *Global Transformation. Politics, Economics and Culture*. Cambridge: Polity.

Javed Burki, S., Perry, G. and Dillinger, W. (1999) *Beyond the Center. Decentralising the State*. (World Bank Latin American and Carribean Studies). Washington DC: World Bank.

Keating, M. (1998) *The New Regionalism in Western Europe*. Cheltenham: Edward Elgar.

Lovering, J (1999) 'Theory led by policy: the inadequacies of the 'new regionalism' (illustrated from the case of Wales)', *International Journal of Urban and Regional Research*, 23: 379-395.

Martin, R. (1993) 'Reviving the economic case for regional policy'. In: RT Harrison and M Hart (eds). *Spatial Policy in a Divided Nation*. Jessica Kinglsey.

Martin, R. (forthcoming) 'Institutionalist approaches in Economic Geography', In: T. Barnes and E Sheppard (Eds.) *Companion to Economic Geography*. Oxford: Blackwell.

Moore, B., Rhodes, J. and Tyler, P. (1986) *The Effects of Government Regional Economic Policy*. London: HMSO.

Myrdal, G (1957). *Economic Theory and Under-developed Regions*. London: Duckworth.

Paxman, J. (1998) *The English*. London: Michael Joseph.

Rodríguez-Pose, A. (1998) *The Dynamics of Growth in Regional Europe. Social and Political Factors*. Oxford: Oxford University Press.

Storper, M. (1998). *The Regional World*. New York: Guilford.

Taylor, P. (1993) The meaning of the North: England's 'foreign country' within, *Political Geography* 12, 136-55.

Thirlwall, A.P. (1980) 'Regional problems as 'balance of payments' problems', Regional Studies, 14: 419-25.

Tomaney, J. (1999a) 'In search of English regionalism: the case of the North East', *Scottish Affairs*, 28: 62-82.

Tomaney, J. (1999b) 'New Labour and the English Question', *The Political Quarterly*, 70, 1: 74-82.

PART I
THE ECONOMY IN TRANSITION

2 Situating the North East in the European Space Economy

DAVID CHARLES AND PAUL BENNEWORTH

Introduction

In the unfolding geography of capitalism, the relative fortunes of places are constantly shifting with the effect that the relative positions and power relations between different regions change (Harvey, 1983; Massey, 1988). New economic spaces are created, some regions experience success, and others decline. Change occurs in tandem with a less dynamic and creative process of general growth, as assets accumulate and their ownership shifts (Schumpeter, 1949). This makes it difficult to evaluate the precise nature of small changes in economies in the relative short-term, whether indicative of a qualitative transformation, or merely an evolutionary development.

For a region such as the North East this is particularly critical in understanding recent developments in the regional economy. The region has already experienced one dramatic qualitative shift. This occurred around the turn of the century, when the industries which had driven the industrial revolution in both the north and the UK as a whole were overtaken by more modern industrial configurations in other regions (Heim, 1985). Elbaum and Lazonick (1986) attributed this to the emergence of corporate capitalism in competitor countries in contrast to the persistence of merchant capitalism in the UK. In the last twenty years there has certainly been a dramatic quantitative shift. There has been a massive growth in service employment concurrently with the withering of the heavy engineering industries (Daniels, 1988). This has been associated with record levels of inward investment across a range of manufacturing and service industries.

A question arises over whether this quantitative transformation is indicative of a deeper qualitative shift in the position of the region. By the 1960s, employment in the region was characterised by branch plants in often nationalised industries (Hudson, 1989a). Yet the nature of the space economy within which the North East is located has undergone dramatic shifts. Industry has experienced restructuring driven by European and global logics, and there has been a sharp increase in the importance and extent of service industries (Amin *et al.*, 1992).

This chapter seeks to address the question of whether there has been a transformation of the region in terms of its relations with other regions within the UK, Europe and the world. The chapter begins by discussing the state of the North East economy in the post-war period, the state from which the supposed metamorphosis has begun. It then presents three perspectives on the current state of the region. The first of these looks at the vernacular discourse which has evolved suggesting that a transformation has taken place. The chapter then compares the changes the region has gone through with other UK and European regions. It is then asked whether in the light of structural changes in systems of economic governance the quantitative changes are themselves significant enough to represent a transformation in the position of the region within Europe. The chapter then takes a third qualitative perspective using examples drawn from sectoral research undertaken within CURDS, and concludes that whilst significant changes have taken place in the region, they are neither as totalising as the vernacular discourse suggests, nor as insubstantial as a quantitative analysis would indicate.

The North East in the national economy, 1945-1975

Recently, some commentators (e.g. Ohmae, 1994) have posited the demise of the nation-state as a key arena of economic and political regulation. However, an examination of changes in relations of production suggests the modern global economy has a dual structure of global networks and reinforced regional communities (Storper and Scott, 1995). Equally important are the relations between the various levels of regulation in response to changing economic structures, the needs of particular industries for co-location and the subsequent geographies of communities of common interest (Scott, 1996).

Thus, although the North East remains critically affected by decisions taken by the national Parliament at Westminster, regulation from the European Union and the impacts of global treaties underpinning the new neo-liberal order have altered the scale around which industries are organised. Similarly, the existence of a few highly successful regional economies masks the fact that the main impact of globalisation has been to increase inter-regional interdependence, hence increasing the significance of national and supranational regulation (Amin and Robins, 1991).

Manufacturing industry - decline and restructuring

The defining feature of the North East economy since the turn of the century has been the decline of its traditional industries.　Various commentators have sought to account for the process of decline.　Elbaum and Lazonick (1986) ascribed indigenous decline to a system of mercantile capitalism that encouraged cost reduction over innovation and the adoption of competitors' best practice.　Heim (1985) argues that poor management was responsible for the decline.　In parallel with this, concentration of industry associated with a national logic of production predicated upon technological advances produced financial concentration around the City of London, rather than the sustenance of independent industrial capital in the peripheral regions.

During the period 1945 to 1975, the reorganisation of industry in the UK evolved around a national logic, facilitated by those prior technological developments, but more importantly, driven by an increased interest in nationalising the industries which had formerly been the economic drivers of the UK, coal, steel and shipbuilding.　Two key developments were to intimately affect the development of the social division of labour in the UK, the geography of high-technology manufacturing industries, and the geography of business and professional services (Massey, 1984).

The geography of manufacturing industry in the UK in the post-war period is intimately tied up with the politics of nationalisation.　Although the term 'nationalisation' in the UK context has acquired exclusive connotations of Keynesian state interventionism, for peripheral regions, there were more important implications for the social division of labour. These were driven by the economic effects of the geographical concentration of sectoral activity in traditional and industrial regions, a process accentuated by central policy hindering the development of new industries in poorer regions.　Nationalisation as a policy was muddled and the worst outcome for the region.　Roberts (1993) argued that in the case of shipbuilding, under-investment prior to nationalisation was sought both by managers, unwilling to bear the costs, and workers, who sought to protect their rigid demarcation.　This pattern of under-investment continued during the period of nationalisation, when control was centralised away from the regions most dependent upon those industries.　Although putatively power lay within a corporatist policy arena, the managers eventual dependence on the government for support and the national organisation of the unions removed control from the regions (Stråth, 1987).

During the dominance of Keynsian statism in the UK, socialisation of old industries was accompanied by the emergence of new industries, which remained largely beyond direct state control. Because of the national logic of economic regulation, these new industries, such as automotive and aerospace industries, emerged in new industrial spaces in the UK, rather than in the established industrial areas. Two main drivers underlay national regulation. Firstly, it was believed new industries in old industrial areas would draw labour away from traditional manufacturing, entirely feasible given the conditions of full cyclical employment of the time (Hudson, 1989a). Secondly, these new industrial spaces were indeed more attractive than old industrial areas for a variety of reasons, related both to their industrial history as well as patterns of government expenditure.

Nationalisation was both a response to the poor competitiveness of individual businesses in common sectors as well as a central policy choice of successive Labour governments. The aggregate effect for regions heavily dependent on traditional industries was to exacerbate these features of their economic structure which independently hindered the development of new modern industries (Hudson, 1989b). That is, old industrialists did not invest sufficiently to revitalise the regions' economics bases, and the workforce were enculturated into the non-entrepreneurial proletariat (Byrne, 1992).

The effect of this strong spatial concentration is illustrated in Table 2.1. As the table shows, those regions which had the highest proportion of manufacturing employment in the modern industries[1] of electrical engineering, aerospace, automotives and electronic capital goods tended to have a much lower employment in the traditional industries of mining, mechanical engineering, metal manufacturing and shipbuilding. The exception to this was Merseyside, whose economy was heavily dependent on the automotive industry, not because it was itself a new industrial space, but because industrial restructuring had reduced to vanishing point the employment in more traditional industries. For the North East, the position was clear. Manufacturing in the region remained reliant on very large externally controlled enterprises, which in turn reduced the attractiveness of the region as a location for new manufacturing industries, and hindered the establishment of new indigenous firms.

Although there was considerable pressure to avoid the establishment of industries which would increase employment competition with traditional industries, there was no such pressure to prevent the establishment of firms employing those outside the sector. There were large numbers of women available for employment at a relatively low cost. Because of their relatively low skill levels, this increased the attractiveness of the peripheral regions for those siting routine manufacturing operations. As the inevitable decline of traditional industries intensified in the 1960s, the attraction of inward investment appeared to offer the potential to create an alternative economic base for the region (Hudson, 1995).

The emergence of the branch-plant economy in the North East is well documented, and by the time of accession of the UK to the EU, a foreign-owned sector (mainly of US origin) of some importance had been established in the North East. Smith and Stone (1989) estimated that in 1971, there were 24,400 jobs in foreign firms, rising to 53,000 in 1978, at a time when total employment in the region was at a level of approximately 1 million. These jobs were predominantly in manufacturing, and Hudson (1995) estimates that 75 per cent of the jobs created were in six sectors: - chemicals, mechanical engineering, electrical engineering, rubber, automotives and printing.

Service industries - new activities, same division of labour

The geography of the development of the service sector in the UK in the post-war period demonstrates a spatial division of labour as marked as that in manufacturing. Government policies had a strong influence on the patterning of employment in the service sector. Organisation of manufacturing around a national logic in turn created pressure for the organisation of business services around a parallel national pattern, and the strength of the City of London ensured that as service industries grew in importance, essential business services were concentrated in London and later the South East (Allen, 1992). Business services were paralleled in the South by a growth in professional services driven by central government expenditure especially in research-intensive fields.

Although there was considerable development in the service sector in the UK in the period 1945-1975, the effects of this were highly unevenly distributed. Table 2.2 illustrates the distribution of employment in the Government Office regions between the extractive, manufacturing and service sectors, illustrating the strong core/periphery split.

However, categorising employment under the broad term 'services' serves to disguise a fundamental feature of the new service space economy of the post-war period. There are two basic forms of service (Kirby, 1995);

knowledge-based services and consumer services. Marshall *et al.* (1988) argue that it is those knowledge-based skills that are the main creators of wealth because they contribute to the addition of value by other firms. The current spatial division of labour is based upon the concentration of the high value-added business services in the core regions, especially London (Kirby, 1995). Table 2.2 would suggest that Merseyside and the South West had by 1975 become centres of service industries comparable to the South East. However, as Table 2.3 shows, producer services by this time were highly concentrated in London, and to a lesser extent, the South East. The peripheral regions, by contrast, tended to have a much higher concentration of consumer services such as retail and construction, as well as higher employment in public administration, education, health and caring. The effects of these drivers was to reinforce a strong division of labour between the old and the new industrial regions, as well as contrasting patterns of service employment between the new service core, and the service periphery.

The state of the region in 1975

Within the logic of a national system of regulation, the position of the North East in the mid-1970s was clear, a peripheral economy dominated by externally controlled businesses, the public sector and consumer services. The fragility of this position was exposed when over the course of the next decade, some 15 per cent of all employment disappeared, whilst the reliance on employment in the foreign-owned sector more than doubled. Headquarters are becoming increasingly responsible for the co-ordinating and strategic functions of businesses; hence, regions with headquarters tend to have higher demand for sub-contracted business services, whilst the manufacturing activities in peripheral regions, although notionally autonomous, have much less demand for such high order skills (Aksoy and Marshall, 1992).

   The next section examines the current position of the North East today from three perspectives. The first of these is from the perspective accepts that that there has been a successful modernisation of industry in the region. This view is held by a particular community within the North East, including elements of the local media, and represents a vernacular discourse.

The North East today - the new orthodoxy

In constructing a vernacular economic-historic discourse for the region, it was popular until very recently to assert that the region had overcome the problems which had persisted since the turn of the century, and successfully achieved a massive historic modernisation of the economy (Robinson, 1992). This modernisation has taken the form of increased employment in more modern sectors, and the continued rise in importance of the service sector. This discourse is produced and reproduced by a community of shared interests in the local media and political and developmental institutions who have an interest in the projection of an image of a vibrant regional economy.

However, recent events have demonstrated that the structural weaknesses of the economy of the North East continue to hamper economic development and reproduce existing structural weaknesses. The volatility of the success of the region and its dependence on external conditions have been demonstrated by recent events concerning two inward investment projects destined for the North East. These projects were instead won by Wales, contrary to the expectations of the local developmental community (Benneworth, 1998).

Yet, the message being reported in both the regional press and by regional institutions until very recently was of an improving region, whose institutional arrangement was admired across Europe. However, there was some confusion evident in the meaning of modernisation in the vernacular discourse. On the one hand, modernisation can mean the upgrading of the position of the region within a core/periphery division of labour. However, in the vernacular discourse, the success of modernisation was evaluated on the basis of the ability of the region to *reproduce its poor position* within that division of labour. Typically, success is equated in the local media as the latter: -'For years, the North East has been streets ahead when it comes to attracting inward investment into Britain' (Northern Echo, 16 March 1997).

Lovering (1999) argues a similar vernacular discourse exists in Wales, which he terms a 'new orthodoxy.' In this new orthodoxy, it is regularly asserted that the problems of a Victorian heritage Wales faces have been overcome through a process of modernisation driven by inward investment. The similarities with the North East are striking; both regions have similar GDP and unemployment levels, significantly worse than the UK average. Indeed, similar criticisms of the vernacular discourse in the North East are equally as valid, with similar interest communities, such as politicians, journalists and regional development practitioners all having an interest in propagation of the welcome news.

Clearly, the North East economy has undergone a significant transformation in the two decades since the UK joined the EU. However, the question this chapter seeks to address is whether the changes in economic structure are indicative of a fundamental change in the position within the division of labour, or whether the changes in the region are a response to a change of scale of the level of regulation to which the regional economy is subjected. The next section therefore considers how membership of the European Union has changed the underlying logic of economic co-ordination.

The European Union and the new logic of production

When the UK joined the EU in 1973, there were no apparent reasons why the European economy would ever be regulated as anything other than a collection of sovereign states with independent economies. Despite the Treaty of Rome creating a customs union in 1957, by 1975 there was no drive to internationalise production. Indeed, many Member States used a combination of monopolies, opacity and favouritism in procurement to ensure that 'national champions' retained systematic advantages in their economies. The chapter now examines the economic changes which ultimately altered the system of industrial regulation within the EU, and the specific effects that this had on the North East.

Over the course of the 1970s, it became clear that the post-war European economic boom was petering out. The early 1970s were a period during which a number of geo-economic changes acted in concert to reduce the competitiveness of European industries with respect to their Japanese and American counterparts (Preston, 1994). Most significant was that growth rates across the EU fell below the levels of the golden decade of the 1950s, and were lower than major partners. Holland (1993) argued that a number of structural problems internal to the EU, in combination with the oil shocks of 1970 and 1973, undermined EU competitiveness much more than either the US or Japan. Labour immobility created spatially concentrated inflationary pressures, and hindered technological adaptation as workers released by investment in new technologies were not easily absorbed elsewhere in the economy. Thus, there was a general concentration of high-technology industries in core regions, and a gradual decline of peripheral regions, in turn stoking inflationary pressures and hastening the onset of the downturn.

However, it is important to bear in mind the external geo-economic changes which undermined the stability of the earlier order as well as the particularity of the EU situation. Firstly, the collapse of the Bretton-Woods currency system opened up the potential for competitive devaluations in the face of declining competitiveness. Although a reasonable response in the short-term, this increased inflationary pressures, which were sufficient to undermine the non-inflationary nature of growth in the post-war period. In the US and Japan, not only were labour markets far more flexible (for cultural reasons), but also both countries sustained domestic demand rather than devalue extensively, reducing the damaging effects of a combination of inflation and stagnation.

The decline in competitiveness of European industry at this time has been referred to by a number of commentators as 'Eurosclerosis' (Preston, 1994). Eurosclerosis can be defined as falling industrial competitiveness with respect to the US and Japan at a time of increasing globalisation of production. This manifested itself in both high levels of inflation and rising unemployment, in contrast to the previously expected trade-off between unemployment and inflation. Delors (1988) argued that there were three possible courses open to the European countries at the end of the 1970s. The first, protectionism was rejected for being what Delors termed 'isolation of the sick', and cost-reduction was rejected because of the collapse in aggregate domestic demand that would arise from the projected necessary 30 per cent cuts in costs through wage cuts to achieve global competitiveness.

Delors argued that the best path to follow was the third, raising the efficiency of European companies. These companies could then increase their productivity through capital investment programmes, and raise the wages of those still employed to sustain domestic demand within the EU. Critical to the North East, Delors admitted that this would necessarily benefit richer countries at the expense of the poor, because of the spatial heterogeneity of unemployment that efficiency drives would precipitate. Regardless of the spatially differential impact of the policy, there was an attempt to create what Delors termed 'Eurogiants,' industries based in Europe, but large enough to compete on equal terms with Japanese and American firms.

The policy of efficiency went through two distinct phases, each of which was associated with a separate regulatory regime, and both of which had significant impacts on peripheral regions. The first phase saw the multi-nationalisation of national firms in a number of sectors, notably pharmaceuticals, chemicals, aerospace and electronics, in which the Commission believed large efficiency gains could be achieved through rationalisation with less danger of reduction of competition (CEC, 1989). Because of the persistence of national forms of regulation the fastest route to achieve efficiency was the creation of firms with a presence in each country in which they operated. In the telecommunications industry, for example, Alcatel built up a multi-national network of firms, each controlled by the central French company but operating separately in each market (Charles, 1994).

The second phase of concentration arrived much later, although the mechanisms for its inception originated in a much earlier phase. A meeting of the European Council in 1985, following the arrival of a new Commission, prioritised both a move to a Single European Market by 1993, and a focus on high-technology growth. The European legislation to implement this took the form of the Single European Act. The SEA was passed in 1986, and permitted an internationalisation of firms organised around a multi-national logic. The SEA removed the right of national governments to procure exclusively from domestic suppliers. In addition to this, it began a process of harmonisation of standards which prevented Member States excluding imports on the ground they failed to meet national regulations, which could be framed specifically to produce such exclusion of non-domestic firms.

The effect of the first phase of the efficiency drive was to concentrate ownership in several sectors in a small number of extremely large firms. However, beyond this concentration of ownership, it remained for the Single European Act, and the regulation of industrial competition at a pan-European level, for the full effects of such concentration to emerge. The elimination of non-tariff barriers to trade removed the necessity of multi-national organisations based around national divisions. By permitting specialisation of production around product lines rather than national markets, this raised the possibility of a pan-European division of labour. The shift from multi-national to international organisation had the effect of further concentrating control at product centres, rather than providing each country with a relatively autonomous headquarters function.

Because of the concentration of control in a few firms located in a relatively small number of European regions, the effects of the rationalisation that such a re-organisation entailed were necessarily heterogenously distributed.   The presence of headquarters or R&D functions increased the attractiveness of those locations to the durability of those functions. This was exacerbated by a particular corporatist form of technology policy which channelled funds through the very largest firms, further concentrating R&D functions in those richest regions lucky enough to host a firm included in the corporatist arena (Sharp, 1990; Charles and Howells, 1992).

However, for those functions, predominantly in mature production, in which cost was a major consideration, the regions of Europe found themselves in competition with other regions to lower their costs and hence increase their attractiveness. This had the general effect of accentuating the divisions between the core and the periphery of the EU. Core regions were in a position to attract the investment to sustain their advantages, whilst peripheral regions aspiring to core status faced the contradiction of raising their labour costs by doing so and making themselves less attractive for their existing industrial base. Certainly, peripheral regions were reliant on the whim of large external companies to make one-off modern investments on the basis of cost considerations. This had the effect of accentuating the division between regions with some element of local control, the core regions, and regions lacking such control, the peripheral regions.

For the North East, this analysis clarifies the current situation. In seeking to answer whether the North East has changed its position within the division of labour, it is necessary to differentiate between quantitative changes arising from the changing spatial scale of production, and qualitative changes in the economic structure that have arisen from increasing the number of functions within the region which are considered as being those provided by the core. The next section looks at a number of quantitative indicators of regional performance to assess what can be inferred from them about the success of the North East within Europe.

The North East in Europe - a quantitative perspective

It is very difficult to directly measure the position of the North East within a European division of labour, especially given the other changes which the economies of Europe have undergone recently. It is therefore necessary to infer the changing position by examining economic changes relative to other regions within Europe. This section looks at a number of economic indicators, such as GDP, R&D and employment structure.

Regional GDP

The fundamental measure of economic success is the relative wealth of a particular region. The North East has for a long time suffered from relative poverty in comparison to the rest of the UK because of its peripheral position in terms of both manufacturing and service industries. Therefore, one would expect, if the economic position of the region were to be improving, that wealth relative to the EU average would similarly be increasing.

However, this is palpably not the case. In 1975 the North had a GDP of 86 per cent of the EU average with some 20 per cent of people living in regions poorer than the North.[2] This position has not improved since 1975, and, currently, there are far fewer citizens of the original 9 Member States present at the accession of the UK resident in regions poorer than the North. Whilst this ignores intra-regional discrepancies, it does suggest that in terms of those 9 Member States, the position of the North has not improved.

It could be the case that because of the continually enlarging European economic space, a much fairer comparison to draw is with all the current members of the EU. However, there is no evidence to suggest that the accession of three very poor countries has improved the relative position of the North; indeed it can be argued that the reverse is true, with the presence of more poor regions placing increased competitive pressures on the North East.

In order to gauge how the region is performing against other regions, this section presents a comparative analysis. In summary, we have chosen three sets of regions or comparators, to offer three sets of characteristic. The first are those regions which have a similar industrial history to the North East, and hence are now facing similar problems. Where data is available, this set includes four regions, Nord-Pas-de-Calais, Pais Vasco, Hainaut and Nordrhein-Westfalen.

The second set are those current core regions whose performance has been exemplary in the last fifteen years. The first of those is Emilia-Romagna, which has succeeded because of vibrant networks of small firms in a variety of industries. The other two regions, Baden-Württemburg and Rhône-Alpes, are both home to a number of very large firms in high-technology industries, with supporting competitive small firms, such as chemicals and mechanical engineering.

The third set are those regions which are in direct competition with the North East for the attraction of inward investment to raise their position within the European division of labour. The regions chosen are Ireland, Scotland and Provence-Alpes-Côtes-d'Azur.

Examining the GDP of these regions suggests that there is no necessary economic determinacy suggesting that declining industrial regions need be poor. Table 2.4 shows the GDP levels of the selected comparator regions in 1994, corrected for currency imbalances and variations in the cost of living. The table shows that the exemplar regions tend to be a great deal wealthier than the declining industrial regions. However, Nordrhein-Westfalen has managed to sustain a high level of wealth despite having a large traditional industrial sector.

Certainly in terms of GDP measures, there is no evidence to suggest that in the last twenty years the North has become relatively richer within Europe, nor that the wealth of the North East is comparable to the core regions within Europe.

Regional unemployment

A second measure of success of regions is their relative unemployment. The European Statistical Office conducts a quarterly Labour Force Survey, measuring unemployment to the ILO standard, which has the effect of harmonising for particular national welfare regimes. Table 2.5 shows the levels of unemployment for the various comparator regions. There clearly is some measure of national regulation on national unemployment rates, determined by national conditions of labour market flexibility. However, the exemplar regions have lower unemployment than either the competitor or the comparator regions.

This data suggests that the North has made some progress in improving its position within Europe. The very high unemployment levels of the 1980s appear to have been systematically reduced, and this would suggest that the region has been successful. However, there are a number of cautionary points worth making. The first is that in the period for which data is available, continental Europe was entering recession, whilst since 1993, the UK economy has been growing. Secondly, employment is not itself an indicator of improved positioning in an international division of labour.

What the analysis does suggest is that the region has been successful in building some competitive advantage. Competitive advantages can take a number of forms, technological, quality or cost. If the cost competitiveness of the North East has increased at the expense of technological and quality competitiveness, then this would indeed be a retrograde step for the region. Indeed, Elbaum and Lazonick (1986) argued that it was such a drive for cost competitiveness over technological advantages which underlay the decline of heavy engineering in the UK at the turn of the century.

The two measures offered above are both problematic. They only measure current success and, in terms of improving the position of the region in an international division of labour, they do not provide unambiguous signals about the direction of the North East. GDP is itself very much a broad and crude measure of economic success, and measures the outcome rather than indicating whether the foundations for success are being development within the region. Similarly, unemployment data suggests that whilst some form of competitiveness has been developed, this could be indicative of a deterioration as much as of an improvement in the division of labour.

Regional capital investment

A complementary approach is to examine the features of core regions, and assess how similar the North East and other comparator regions are to them. One particular measure of the core regions is a very high level of capital investment per employee, given the tendency of the core to host R&D and management functions as well as the latest and most innovative manufacturing functions, which in turn require high capital investment.

Table 2.6 shows the investment in fixed capital by region in 1992. Although the UK figures were depressed by a recession, more recent data suggests that even at the depths of the slump on the continent, poorer regions were sustaining investment at levels systematically higher than in the North of England. In 1994, for example, investment in Emilia-Romagna was 6088 ECU per employee in comparison to 2820 ECU per employee in the North.

This is a more worrying indicator, suggesting that in the future the position of the region will decline relative to other regions in which much higher levels of investment are being made. However, because there are a number of poorer regions in which investment levels have been high for a long period of time without discernibly improving their position, it may not be necessary to infer a bleak future for the North from the investment data.

Regional investment in R&D

Another feature of core regions is high levels of investment in research and development functions. Indeed, such functions are central to core regions, providing the innovations through which the competitive advantage of these otherwise very high-cost locations is sustained. An exception to this rule is Emilia-Romagna, because of the particular markets it operates in, and its organisation around very small firms (Storper, 1993). In Emilia-Romagna much of the innovation is informal in nature, and hence not recorded in official figures. In contrast to this situation, Baden-Württemburg and Rhône-Alpes are both homes to a number of very large world-class firms with large dedicated R&D facilities, such as Rhône-Poulenc, BASF and Mannesmann.

What is clear is that the particular national system of innovation is very important to determining national levels of investment in R&D. The German system of innovation is rooted in the Prussian centralist-modernising state of the late nineteenth century, under which large industrial combines were built up, large numbers of polytechnics established and engineering formally professionalised and valorised. Thus the level of R&D in Nordrhein-Westfalen is much higher than in non-German declining industrial regions. This is because of the presence of very large and technically competent firms as well as a technically-focused education and research system which continue to support the German industrial structure to this day (Keck, 1993).

The North East appears to perform very badly in terms of investment in R&D. At this point it is worth noting that the figures for the North, first produced in 1992, suffered from mis-recording which tended to emphasis R&D spending at the firm headquarters. Since 1993, the (corrected) figure for the North East has converged at 1.26 per cent, a figure similar to that of both competitors and comparators (as distinct from exemplars).

The North East is dually constrained by a national system of innovation which centralises innovation in the South East (core) whilst undervaluing the engineering profession and the importance of innovation (Walker, 1993). Firstly, the North East has almost no government R&D investment, one-eleventh that in the South East,[3] and one-eighth that of the South West (Economic Trends, July 1997). This is predominantly in defence research, which has the effect of crowding out more productive public research as well as having specifically spatially heterogeneous impacts (Morgan, 1986). Similarly, higher education R&D spending in the North East is half that of London, and two-thirds that of Scotland.

Secondly, business R&D investment in the region has suffered from the restructuring of industry at a European scale, with firms such as NEI,

ICI, and Joyce Loebl all either being taken over or internationalising their R&D operations externally because of the much more amenable conditions outside the region. These conditions in the UK are necessarily supported by the national spending regime which concentrates public R&D funds in areas of existing expertise.

Employment in knowledge intensive businesses

Just as R&D investment figures fail to capture informal innovation in small firm networks, as evidenced by data for Emilia-Romagna, so they have become increasingly obsolescent with the increased importance of the service economy. As outlined above, there are two main categories of service, routine consumer services, and much higher-value producer and professional services. Table 2.7 also shows an index for employment in higher level producer services for the UK Government Office Regions (no directly comparable European data is available). The table shows that there is a clear core-periphery split in the levels of service present in each of the regions.

The North East has increased proportional employment in service industries over the last twenty years, from 51.9 per cent to 66.2 per cent (1975-1995). However, in that period, total employment in service industries has increased by a mere 5 per cent across that time period. Critical to the region is the fact that much of the employment thus created has been in low- and unskilled positions, such as retailing and call centres. Particularly with call centres, such positions are often vaunted in popular discourse as high-technology, but the high technology is embodied in smart systems which are designed in, and imported from, outside the region. Moreover, Richardson and Marshall (1996) argue that for one of the factors affecting relocation decisions to Tyne and Wear is the low cost of the workforce.

Another factor reducing the number of high-level service skills in the North East is that many firms have their mid-level business service needs met by businesses in Leeds and Birmingham rather than within the North East (Robinson, 1988; Jones 1997a). The jobs that have been created have a low potential for the addition of value in the region, being predominantly in tourism, leisure, consumption and sales.

For the North East, there has been a shift towards the importance of the service industry, but this is not necessarily indicative of a change in the relative position of the region. Every region within the European Union has experienced a large increase in service employment in the last twenty years. The current state of service employment in the region appears to be dominated by public sector employment in health, education and administration, and service-sector analogues of the branch-plants which have dominated manufacturing in the region since the decline of heavy engineering after the 1960s.

Just as the manufacturing branch-plants failed to provide a stable economic base for the region, being subject to periodic disinvestment (Hudson, 1995), these service branch-plants, back offices and call centres are similarly vulnerable to disinvestment. Although the region appears to have a number of advantages over other regions, such as a low cost base, and high education levels, there are very few multiplier effects from such businesses. They also contribute relatively little to the development of regional skills which increase the attractiveness of the region for new investors.

Inward investment

The North Eastern economy is now highly dependent on foreign-owned businesses, many of whom are attracted by low production costs and government incentives to locate in the region. The foreign firms do not contribute excessively to the value added in the region. Table 2.8 shows that although the most peripheral regions of the UK are most dependent on foreign firms for value addition, the South East enjoys more value added by foreign firms than the North East. Indeed, given the structure of the UK economy alluded to by Walker (1993), a higher value might indicate a higher degree of internationalisation rather than merely a higher degree of external control.

What is significant about foreign investment in manufacturing in the North East is the contribution it makes to the total investment in the region. The four most geographically peripheral regions of the UK, Northern Ireland, Scotland, Wales and the North East, all have approximately half their investment in manufacturing contributed by foreign-owned enterprises, whilst such investments produce only one-third of gross value added. This suggests that in peripheral regions, foreign firms are investing considerably more than indigenous firms. This in turn suggests that the indigenous capacity of the peripheral regions are much lower than the core regions, and thus the gap between them is not being reduced.

The North East is in a position of reliance upon foreign firms to make necessary investments in modernisation of the regional economy, because of the poor performance of not just local firms but UK firms in the region. Reliance on foreign investment is problematic because of the short duration of the investment, and the propensity of firms in the North East to shed employment (Collis, 1992). Investments are typically made for a short period, without guaranteeing survival or subsequent reinvestment. Over-reliance on inward investment is therefore inhibiting the development of a continuing virtuous dynamic of production and reinvestment.

The position of the North East

It appears that the North East has failed significantly to change it its position within the spatial division of labour in the last twenty years. Although the composition of industries has changed, and there has been a shift from manufacturing to service employment, other changes have failed to occur. There has not been an increase in R&D in manufacturing, nor in the most entrepreneurial service occupations. A service branch-plant sector has emerged to complement a burgeoning foreign-owned manufacturing sector. The much heralded explosion of far eastern investment has been paralleled by a much less visible process of disinvestment by UK and US-owned firms. In effect, the evidence presented above appears to suggest that the North East is in no better a position than it was twenty years ago. However, it is clear that the region has achieved some successes. The region has built from almost nothing in a decade an efficient and high-quality automotive sector (admittedly based around one foreign-owned firm). The region has produced a number of high-technology firms near enough to the cutting edge of technology to be bought out by very large supernational firms, such as Sun and Viasystems. The region has managed to support two firms producing accountancy software for a global market, QSP and Sage, which have both undergone extensive overseas expansion whilst retaining managerial autonomy within the region. Indeed, Sage recently bought out an American software developer, State Of The Art, to complement its US operations. It is therefore necessary to avoid a monotonic view of the region as a failure in terms of upgrading its position within an international division of labour. There is a more heterogenous pattern to the success of the region, reflected to some extent in widening income inequalities in the region, and the growth of the working poor. The next section seeks to refine the analysis presented above by complementing the quantitative presentation with a qualitative description drawing more heavily on studies of specific sectors.

The North East - global node or outpost

It is clear that something has changed; industrial structure is very different, and the critical question remains whether this is indicative of a qualitative change in the fundamentals of the North East economy, or merely the new manifestation of a branch-plant economy in the 1990s. This section attempts to portray the problematic of regional analysis given a heterogenous and divergent socio-economic situation. This is represented by the co-existence of branch-plants with global firms, and the existence of a new class of business not present in the 1960s, the so-called 'glocal' firm, which changes the way the division of labour is conceptualised (Charles and Benneworth, 1997). In evaluating the performance of the North East, it is perhaps disingenuous to assume too much similarity between the North East and other European regions. The situation in the North East prior to 1975 was not only of a region with declining industries, but a region with serious structural problems. In particular, the industries in the region, in common with the rest of the UK, were organised around a peculiarly British form of mercantile capitalism which engendered short-termism and cost reduction over innovation, technological development and long-term growth. To have overcome that particularly British problem is an accomplishment prerequisite to any kind of change in the region's position within a pan-European division of labour.

Manufacturing remains relatively important to the region, and various kinds of firms have contributed to the introduction of new products, practices and processes to the region. The willingness of the region to adopt adaptation over confrontation may be seen as a process of the intensification of work, accompanied by social fragmentation with the aim of preserving the existing regime of accumulation in a new right-wing climate (Hudson, 1989b). However, some of the imported processes, practices and products introduced through the diffusion of casualised Fordist industry (both in manufacturing and services) have driven an important modernisation of the economy, without which, structural problems of the region would have limited any demand-sided attempts to increase the position of the region with the division of labour.

This section seeks to reconcile the fact that despite the region still being a branch-plant economy, the particular form of this economy has changed due to a process of internationalisation. In one sense, as the recent Asian currency crisis has demonstrated, the branch-plants remain highly volatile - the troubled Korean company Samsung, recently cancelled a proposed 1,200 job expansion to the electronics manufacturing business at Wynard, Teesside. However, investments such as Viasystems at Balliol offer employment resembling that available through heavy industry in the past, although with a greater emphasis on the integration of innovation into the manufacturing process, and the adoption of the corporate capitalist form of organisation.

There is a great variety of forms of economic activity in the region, and much potential to raise economic fortunes. This section examines how the performance of the economy is in some areas creating indigenous potential, with the capacity to add considerable value in the region. More problematic is the extent to which these factors are general within the region's economy, or concentrated in small areas. Certainly inequalities in the region have increased, in common with the rest of the developed world, over the last twenty years. Thus, this section concludes by examining how heterogeneous the unfolding of those processes has been within the North East.

Sustaining employment

It is becoming increasingly apparent that concurrent with the end of lifetime employment has arrived the end of the life-time career. This means that individuals need to continually be upgrading and developing their skills to deal with technological changes that affect their career trajectories. For younger people, this can and has been done through modernisation of the education and training system. Modern apprenticeships have been designed to permit employers to work in partnership with further education colleges to tailor training to the needs of both the firm and the trainee.

However, for a number of employees in the region, there has been considerable development of their careers through training whilst in employment. There are a number of skills prerequisite to the majority of forms of employment, and low unemployment, as well as being socially desirable, helps to sustain the levels of these skills in the region. One electronics firm, for example, has introduced fully automated BACS wage payment, which requires employees operate a current account. Although it may appear superficial, anecdotal evidence suggests that in some marginalised areas the lack of capitalist interaction skills, which those raised in a culture of employment take for granted, are a barrier to their employment.

Although employment in cost-minimising and under-investing firms can hurt the regional economy, there is some evidence to suggest that not all firms in the region behave that way. In the electronics industry, for example, 29 per cent of firms put a significant effort into training, covering 38 per cent of a workforce of 10,000 in the region (Charles *et al.*, 1997). The occupational structure in the region is not too dissimilar to other regions in the UK, and if problems exist it is that there are too many employed on unskilled occupations, for whom training can offer few opportunities and, the consensus suggests, whose jobs stand to diminish continually over time.   There are fewer in managerial, technical and professional occupations, those most prone to entrepreneurial activity. Table 2.9 shows the proportion of economically active by occupation in 1996 for a selection of regions of the UK, which suggests that certainly not all employment in the North East is in predatory and exploitative competitive industry.

The North certainly faces a problem of a mismatch between the skills of the unemployed and the potential employment the region can offer (Jones, 1997b). Some 20 per cent of unskilled workers in the North East are unemployed compared with a national average of 10 per cent, and proportionally fewer managers in the North East are unemployed compared to the national average (Employment Gazette, 1996; Labour Force Survey, 1997). Thus, by increasing employment which potentially leads to raising individual employability, the performance of the region is being increased. A key need of the region is therefore to provide conduits for the unemployed to enter successfully into employment.

Modernisation of the industrial base

Given historic under-investment in the industries of the region, and the decline of the traditional industries through which regional development had previously been driven, without substantial government assistance, the North East lacked both a modern industrial base and the means through which it could be upgraded.   Foley and Macaleese (1991) argued that Ireland faced similar problems at a similar time, and wholeheartedly embraced upgrading through inward investment. Despite not being the best option, it was the most efficient and realistic option given the lack of wherewithal to publicly fund such a modernisation.

Dunning first raised the importance of organisation and business culture, when he noted that US car firms manufacturing in the UK had a much higher productivity than British firms (1958). Because of the mercantile organisation of much of the heavy engineering in the region, industry was effectively isolated from developing best practices, which exacerbated the competitiveness problems those industries felt. The influx of new industries to the region has not merely modernised the capital stock within the region, but the cultures of production within the region. The presence of firms inculcating a modern employment culture as well as developing their staff is important in sustaining the position of the region, providing at least a platform for future developments.

Connections to global networks

As important as the import of new cultures to the region is the presence of regional actors within global networks. The analysis outlined above assumes that new cultures are transmitted to the region through the arrival of new firms with their own particular management styles developed overseas. However, such a process is necessarily lumpy; a much more efficient way of sustaining management advantages is through participation of key actors in international knowledge and learning networks, through which best practice can be disseminated. These networks operate both formally, and informally. In a formal sense, firms network through material transactions, and the less routine the transaction, the greater the need for physical proximity or stronger networking to overcome the problems arising in the transaction process.

There are several types of firms in the North East actively participating in formal global networks. Regional multi-nationals are either generating best practice or adapting their performance to that recognised by the parent company. There are those firms who supply to regional multi-nationals, who continually have to adapt their own practices to the needs of the original equipment manufacturer; this category includes both clusters around Nissan and ICI. There are firms who export beyond the region and the UK, and who innovate in response to continual contact with their user base. There is also a certain degree of networking between firms and universities in the region. All the universities in the North East are involved in the provision of knowledge to the business sector through the Knowledge House initiative. Finally, some firms in the region are successful local firms which have been taken over by multinationals. As well as being indicative of the dormant potential of the region to generate successful indigenous companies, they provide a means through which small local companies can have ready access to global financial, technical and professional networks.

Informal networks exist where there are non-material transactions between firms, such as flows of staff or ideas. These networks may exist because of physical proximity, through occasional meetings, or through either real-time or recorded media. There is a steady flow of staff between firms in the North East and beyond, which permits a steady flow of expertise to the region embodied in the staff - given the small size of the regional economy, for particularly specialist tasks, this is an important mode of recruitment and hence becomes an important conduit for information. Firms may come together to share their experiences, to learn and to teach in a number of arenas. Thus even small local firms may share in what is effectively global expertise. If such networks have academic input, then this extends the range of technical advice which can enter the regional network.

World class firms

Most significant for the evolution of the regional economic base is the presence of a number of locally headquartered export-intensive and multi-national firms. There are very few of these companies, but they span both traditional (e.g. Domnic Hunter) and modern sectors of industry (e.g. Sage). As well as suggesting that the position of the region is more ambivalent than the tag 'branch-plant' would suggest, such firms have the potential to have a massive impact on the region in a relatively short period of time, and effect the qualitative changes the region needs as opposed to much slower changes.

Such firms have the potential to increase employment, particularly in those occupational classes of which the North East is short - technical, managerial and professionals. Although it is naive to suggest that such a supply of these individuals will make the region more attractive for firms to locate headquarters, there is certainly a mild improvement in the indigenous capacity, by providing a variety of modes of support for those entrepreneurs who would seek to develop their own successful companies, in terms of managerial expertise, mentoring and the availability of skilled staff.

The problematic of North East regional development

This chapter has presented a number of perspectives on the position of the North East economy. In the vernacular discourse or new orthodoxy, the region has in the recent past been presented as being highly successful merely by those with an interest in such as portrayal. A quantitative analysis suggests that the region still has the characteristics of a poor branch-plant region similar to other similar European regions with common characteristics. A more qualitative analysis suggests that elements of both of these readings of the situation are apparent, and that whilst the region is no longer a branch-plant economy in the 1960s sense, it is fatuous to assert it is within the European economic core, as is frequently (and equally inanely) asserted about Wales.

The first criticism to level at the employment situation in the region is the fact that employment in the UK, even when standardised to the ILO definition, is sufficiently affected by the benefits system to produce under-reporting of the order of 50 per cent (Beatty and Fothergill, 1996; Green, 1997). Thus, although there are few unemployed in the region, there are many non-participants in the regional economy from whom the benefits of the re-industrialisation have been excluded.

The second criticism of the transformation thesis is that the modernisation of the industrial culture has been an increasingly problematic process, both on an individual and a regional level. The rise of casual Fordist production has engendered all of the problems of the traditional industries with few of the benefits. For an individual, work is demanding, intense, and encourages a culture of employment over entrepreneurialism whilst lacking the certainty which Fordist systems of employment regulation offered (Beck and Beck-Gernsheim, 1996). At a regional level, social fragmentation has resulted in contrast to the class solidarity produced by common experiences in traditional industries (Colls, 1992).

The third criticism is that although there are many firms connected to the global economy, these firms represent a very small sector of employment. The public sector and untraded service sector amount to approximately 60 per cent of employment in the North East. Additionally, even in those firms well-connected into global networks, there has been a fragmentation of the workforce between the skilled and unskilled, those in a position to benefit from globalisation and those excluded. Thus although such networked firms can contribute to an upgrading of the region by improving business conditions, it is important not to overstate their direct importance to employment.

Conclusions

The North East has come through a great many changes in the last twenty years. Whilst it is clear that the North East is not a core region within Europe, it is equally incorrect to state that the position of the region within a European division of labour is merely a new manifestation of a branch-plant region in the age of globalisation. There have been two changes within the regional economy with effects too significant to categorise as merely incremental or quantitative.

The first is the elimination of the traditional engineering base of the region. In some senses this was disastrous, producing mass male unemployment, reducing the capital stock and reducing the demand for high order engineering and professional services within the region. However, and more contentiously, some commentators suggest that those industries had proved themselves unable to modernise their practices, and so their disappearance was part of a process of modernisation, which has led to the introduction of new processes and practices which have enabled the region to compete more successfully in new markets. The key issue is whether such an asymptotic modernisation process was a prerequisite to a new phase of strong industrial growth, or the product of systematic tri-partite of the regional economy.

On one hand, the evidence suggests that the position of the region has not significantly declined since 1975 - regional GDP is still about the same compared to the EU average as in 1975, when the region was dominated by traditional industrial employment. More speculatively, other regions within Europe may have to undergo this restructuring process. However, it could be argued that the region lacks the necessary industrial capacity to sustain this position into the future. Similarly, traditional industries in other regions may well be better poised to survive because of the higher levels of investment they have enjoyed over time.

The second change has been that now much industrial capacity in the private sector is externally owned - there are many locally-owned business, but they employ relatively few people. This necessarily reduces the benefits which accrue to the region from economic activity and, because of the way research and investment decisions are taken by externally-owned firms, may stunt the future development of the region. The external ownership of assets increases the irregularity of the development process, especially given increased competition the region faces for inwards investment.

However, it has already been asserted that the development process in the region was seriously hindered by the organisation of capitalism in the UK which militated against the formation of regionalised industrial capital to the benefit of centralised financial capital. Given the relatively high levels which foreign companies are investing in the region relative to their generated input, it suggests that indigenous companies may still be hindered in such a manner, and reliance on (weak) local capacity may run the risks of reproducing the fault lines ultimately responsible for the collapse of traditional industry in the region.

The key issue for the region is whether modernisation has been a process of disciplining of labour, as Hudson argues, or an essential prerequisite to future growth. Of course, the two points are not necessarily mutually exclusive, and there need be no strong correlation between the intentions of actors who took the decisions affecting the development trajectory, and the final appearance of the economy situation.

What is critical to the success of the region is that the necessary policy framework is implemented taking cognisance of the current situation. Such a policy framework would encourage the development of those elements which would engender a dual evolutionary track of high-order business functions and more locally-owned companies. It is anticipated that the main barrier to this will be the traditional concern of British politicians with employment generation at any cost (McKay and Cox, 1979).

A period of great qualitative change has been experienced in the North East economy - there is little traditional industry remaining to disappear, nor can the proportion of foreign ownership increase more significantly than producing one-quarter of manufacturing output. Certainly, the process of change has been highly socially unjust, generating much opposition. Conversely, certain vernacular communities have been hasty in proclaiming its success, creating a cynicism in those outside that community. Substantive efforts to improve the embedding of high-order functions in the region can use the modernisation as a foundation. What is clear is that the region is still not in a position to succeed on its own account; any substantive success lies in the future. The region may appear to have stood still for two decades, but it has come a long way in that time.

Notes

[1] 'Modern industries' refers to automotives, aerospace, instrument/electrical engineering and electronic capital goods; 'traditional industries' refers to mining, mechanical engineering, shipbuilding and metal manufacturing.

[2] GDP data is only available for the North region, comprising the North East and Cumbria. The North East figure will be slightly lower as GDP per head in Cumbria is higher than the rest of the region.

[3] Government Office South East region, surrounding the GO London region.

References

Allen, J. (1992) 'Services and the UK space economy: regionalisation and economic dislocation', *Transactions of the Institute of British Geographers,* 17, 123-142.

Aksoy, A. and Marshall, N. (1992) 'The Changing Corporate Head Office and its Spatial Implications', *Regional Studies,* 26, 149-162.

Amin, A., Charles, D. and Howells, J. (1992) 'Corporate restructuring and cohesion in the new Europe', *Regional Studies,* 26, 319-332.

Amin, A., and Robins K. (1991) 'These are not Marshallian times', in R. Camagni (ed)., *Innovation networks: spatial perspectives,* London: Belhaven.

Beatty, C. and Fothergill, S. (1996) 'Labor-market adjustment in areas of chronic industrial decline - the case of the UK coalfields', *Regional Studies,* 30(7) pp.627-640.

Beck, U, and Beck-Gernsheim, E., (1996) 'Individualisation and "precarious freedoms": perspectives and controversies of a subject-oriented sociology' in P. Heelas, S. Lash and P. Morris (eds) *De-traditionalisation: critical reflections on authority and identity at a time of uncertainty,* Oxford: Blackwell.

Benneworth, P.S. (1998) 'Comparative institutional constraints on expressions of regional identity - institutionally-constructed regional identities in declining UK industrial regions' *Linking theory and practice: issues in the politics of identity,* September 9-11, 1998, University of Wales, Aberystwyth, UK.

Byrne, D. (1992) 'What sort of future?' in R. Colls and B. Lancaster, (eds) *Geordies: roots of regionalism,* Edinburgh: Edinburgh University Press.

Charles, D. (1994) 'Alcatel: a European champion for a globalising market', in J.-E. Nilsson, P. Dicken and J. Peck, *The internationalisation process: European firms in global competition,* London: Paul Chapman.

Charles, D.R. and Benneworth, P.S. (1997), 'Electronics in the North East', *Northern Economic Review,* 26, pp. 32-46.

Charles, D. and Howells, J. (1992) *Technology transfer in Europe: public and private networks* London: Pinter.

Charles, D.R, Naylor, J.R. and Benneworth, P.S. (1997) 'Electronics in the North East' *Report for the North East Labour Market Information Group.*

Collis, C. (1992) 'Overseas inwards investment in the UK regions' in P. Townroe and R. Martin (eds) *Regional development in the 1990s: the British Isles in transition,* London: Jessica Kingsley.

Colls, R. (1992) 'Born again Geordies,' in R. Colls and B. Lancaster, (eds) *Geordies: roots of regionalism,* Edinburgh: Edinburgh University Press. *competition policy,* Luxemburg:

Daniels, P. (1988) 'Producer services and the post-industrial space economy', in D. Massey, and J. Allen (eds). *Uneven redevelopment: cities and regions in transition,* London: Hodder and Stoughton.

Delors, J. (1998) *Our Europe - the Community and National Development,* London: Verso, (trans. 1992).

Dunning, J.H. (1958) *American Investment in British Manufacturing Industry*, London: George Allen and Unwin.

Elbaum, B., and Lazonick, W. (1986) 'An institutional perspective on British decline', in B. Elbaum and W. Lazonick, (eds) *The decline of the British economy*, Oxford: Clarendon.

Foley, A., and McAleese, D. (1991) *Overseas Industry In Ireland*, Dublin: Gill and Macmillan.

Government Statistical Service (1997) *Regional competitiveness indicators: a consultation document*, London: The Stationary Office.

Green, A.E. (1997) 'Exclusion, unemployment and non-employment', *Regional Studies* 31(5) 505-520.

Harvey, D. (1983) *The limits to capital*, Oxford: Blackwell.

Heim, C. E., (1985) 'Interwar responses to regional decline', in Elbaum, B., and Lazonick, W., (eds) *The decline of the British economy*, Oxford: Clarendon

Holland, S. (1993) *The European imperative: economic and social cohesion in the 1990s*, Nottingham: Spokesman.

Hudson, R. (1989a), *Wrecking a region: state policies, party politics and regional change in North East England*, London: Pion.

Hudson, R. (1989b), 'Labour market changes and new forms of work in old industrial regions: maybe flexibility for some but not flexible accumulation', *Environment and Planning D: Society and Space*, 7 (1), 5-30.

Hudson, R. (1995), 'The rôle of foreign inward investment,' in L. Evans, P. Johnson and B. Thomas, *The Northern regional economy: progess and prospects in the North of England*, London: Mansell.

Jones, I. (1997a) 'Professional services in the North of England', Report to the *Services Challenge*.

Jones, I. (1997b) 'Winners and losers in tomorrow's job market', *Northern Economic Review*, 26, pp. 20-31.

Keck, O. (1993) 'The National System for Technical Innovation in Germany', in R. R. Nelson, *National Innovation Systems: A Comparative Analysis*, Oxford: Oxford University Press.

Kirby, P. (1995) The development of the service sector, in L. Evans, P. Johnson & B. Thomas (eds) *The Northern Regional Economy*. London: Mansell.

Lovering, J. (1998) 'Misreading and misleading the Welsh economy; the "new regionalism" *Papers in Planning Research* Department of City and Regional Planning, Cardiff: University of Wales.

Lovering, J. (1999) 'Theory led by policy: The inadequacies of the 'New Regionalism' (illustrated from the case of Wales)', *International Journal of Urban and Regional Research* 23, 379-95.

Marshall, J.N., Wood, P., Daniels, P.W., McKinnon, A., Bachtler, J., Damesick, P., Thrift, N., Gillespie, A., Green, A. and Leyshon, A. (1988) *Services and uneven development*, Oxford: Oxford University Press.

Massey, D. (1984) *Spatial divisions of labour: social structures and the geography of production*, Basingstoke: Macmillan.

Massey, D. (1988) 'Uneven development: social change and spatial divisions of labour', in D. Massey, and J. Allen (eds). *Uneven redevelopment: cities and regions in transition*, London: Hodder and Stoughton.

McKay D.H. and Cox, A.M. (1979) *The politics of urban change,* London: Croom Helm.

Morgan, K. (1986) 'Re-industrialisation in peripheral Britain: state policy, the space economy and industrial innovation', in R. Martin and R. Rowthorn, *The geography of de-industrialisation*, Basingstoke: Macmillan.

Ohmae, K. (1994) The rise of the region-state. *Foreign Affairs* pp.79-87.

Preston, J. (1994) *EU Policy Briefings - Regional Policy*, London: Longmans.

Richardson, R. and Marshall, J.N. (1996) 'The growth of telephone call centres in peripheral areas of Britain: evidence from Tyne and Wear', *Area,* 28(3), pp. 308-317.

Roberts, I. (1993) *Craft, Class and Control: the Sociology of a shipbuilding community*, Edinburgh: Edinburgh University Press.

Robinson, F. (1988) *Post-industrial Tyneside: an economic and social survey of Tyneside in the 1980s*, Newcastle upon Tyne City Library and Arts.

Robinson, F. (1992) 'The Northern region', in P. Townroe and R. Martin (eds). *Regional development in the 1990s: the British Isles in transition*, London: Jessica Kingsley.

Schumpeter, J.A. (1949) *The Theory of Economic Development*, Cambridge MA, Harvard University Press.

Scott, A.J. (1996) Regional motors of the global economy, *Futures* 28 (5), 391-411.

Sharp, M. (1990), 'The Single Market and European Policies for Advanced Technologies', in C. Crouch and D. Marquand, *The Politics of 1992: Beyond the Single European Market*, Oxford: Blackwell.

Smith, I. and Stone, I. (1989) 'Foreign investment in the North: distinguishing fact from hype,' *Northern Economic Review* 18 pp 50-61.

Storper, M. (1993), 'Regional "Worlds" of Production: Learning and Innovation in the Technology Districts of France, Italy and the USA', *Regional Studies* 27 (5) pp. 433-455.

Storper, M. and Scott, A.J. (1995) 'The wealth of regions: market forces and policy imperatives in local and global context', *Futures,* 27 (5), pp. 505-526.

Stråth, B. (1987) *The Politics of de-industrialisation: the contraction of the Western European shipbuilding industry*, Beckenham: Croom Helm.

Walker, W. (1993) 'National Innovation Systems: the United Kingdom', in R. R. Nelson (ed). *National Innovation Systems: A Comparative Analysis*, Oxford: Oxford University Press.

Table 2.1   The percentage of manufacturing employment in 'modern' and 'traditional' industries, 1975

Region	'Modern'	'Traditional'
Eastern	32.6	15.8
S East	30.5	21.5
W Mids	29.4	25.0
S West	27.6	20.9
Merseyside	26.5	12.4
London	26.4	12.8
E Mids	23.1	30.6
N West	19.8	17.1
Wales	18.7	34.9
N East	16.6	36.9
Scotland	16.4	29.0
Yorkshire and Hmside	10.7	26.7%

Source: Census of Employment, 1975 © Crown Copyright Reserved ONS Statistics (NOMIS), 1998.

Table 2.2   The distribution between sectors of employment in the current Government Office regions, 1975 (%)

1975	Primary	Secondary	Tertiary
London	0.5	22.0	73.0
S East	9.9	27.7	64.3
S West	14.4	28.0	61.8
Merseyside	3.2	33.6	60.3
Scotland	13.5	30.4	57.1
Eastern	11.3	33.8	56.8
Wales	21.6	31.7	54.9
E Mids	25.9	32.0	53.9
N West	3.9	40.4	52.7
Yorkshire and Hmside	16.2	36.7	51.8
N East	15.2	35.3	51.5
W Mids	5.7	46.0	46.6

Source: Census of Employment, 1975 © Crown Copyright Reserved ONS Statistics (NOMIS), 1998.

Table 2.3   The composition of employment in service sub-sectors by Government Office region, 1975 (%)

	Professional services	Other private services	Public admin	(of which national govt)	Temporary emplymt as % all emplymt
London	17.60	56.10	26.20	5.50	73.00
S East	11.00	55.60	33.40	5.00	64.43
Eastern	9.80	57.10	33.10	3.10	56.80
N West	8.90	58.40	32.70	3.20	52.70
W Mids	8.80	57.30	33.90	2.80	46.60
S West	8.60	59.20	32.20	5.50	61.80
Scotland	8.50	58.00	33.50	3.90	57.10
Merseyside	8.40	60.00	31.60	3.40	60.30
E Mids	7.30	57.80	34.90	3.00	53.90
York and Hum	7.30	58.70	34.00	2.80	51.80
N East	6.20	59.30	34.50	4.90	51.50
Wales	5.80	57.20	37.10	5.30	54.90

Source: Census of Employment, 1975 © Crown Copyright Reserved ONS Statistics (NOMIS), 1998.

Table 2.4   **GDP at Purchasing Power Standard per capital by EU NUTS I region, 1994**

Region	*Includes*	1994
North (UK)		**14180**
Nord-Pas-de-Calais		14440
Noreste	*Pais Vasco*	14800
Region Wallone	*Hainaut*	15100
Nordrhein-Westpfalen		18620
Emilia-Romagna		21300
Baden-Württemburg		20930
Centre Est	*Rhône-Alpes*	16950
Ireland		14700
Scotland		16350
Mediterranée	*Provence-Alpes-Côtes-d'Azur*	15170

Source: Eurostat (1995), Gross Domestic Product (Regio), Luxemburg Eurostat (producer), **r•cade** online service (distributor), Universities of Durham and Essex; Regional Trends, 1997; Economic Trends, 1997.

**Table 2.5   Harmonised unemployment rate by EU NUTS I region,
1993-1996 (%)**

Region	Includes	1993	1994	1995	1996
North (UK)		**11.5**	**11.4**	**11.0**	**9.5**
Nord-Pas-de-Calais		14.0	16.0	15.4	16.8
Noreste	*Pais Vasco*	19.4	21.5	19.3	17.9
Region Wallone	*Hainaut*	11.2	13.0	12.9	12.9
Nordrhein-Westpfalen		7.0	8.4	8.0	8.4
Baden-Württemburg		4.4	5.7	5.2	5.6
Centre-Est	*Rhône-Alpes*	10.8	11.2	10.2	10.7
Emilia-Romagna		6.3	6.5	6.3	5.3
Ireland		15.7	14.7	12.2	12.4
Scotland		10.1	9.6	8.7	8.0
Mediteranée	*Provence-Alpes-Côtes-d'Azur*	14.6	15.4	14.8	16.1

Source: Eurostat (1995), Harmonised Unemployment (Regio), Luxemburg Eurostat (producer), r•cade online service (distributor), Universities of Durham and Essex; Regional Trends, 1997; Economic Trends, 1997.

Table 2.6   Gross fixed capital formation per employee by EU NUTS I region 1992 (where available)

GFCF per employee (ECU)	*Includes*	1992
North		**2830**
Nord-Pas-De-Calais		8336
Region Wallonne	*Hainaut*	2578
Nordrhein-Westfalen		9029
Baden-Wurttemberg		10576
Centre-Est	*Rhône-Alpes*	8459
Emilia-Romagna		7101
Ireland		4097
Scotland		2518
Mediterranee	*Provence-Alpes-Côtes-d'Azur*	7981

Source: Eurostat (1995), Economic Accounts by Industrial Sector (Regio), Luxemburg Eurostat (producer), r•cade online service (distributor), Universities of Durham and Essex; Regional Trends, 1997; Economic Trends, 1997.

Table 2.7   High level occupation index for the UK standard regions 1992-1997 (UK=100)

	1992	1997
London	125	127
S East	111	110
Eastern	104	102
UK	100	100
S West	99	97
Scotland	92	94
Merseyside	94	93
W Mids	88	93
North West	97	91
YandH	89	91
E Mids	90	91
Wales	93	90
North East	**83**	**83**

Source: Annual Employment Survey. © Crown Copyright Reserved ONS Statistics (NOMIS), 1997.

Table 2.8   Gross value added in manufacturing by foreign companies as a percentage of all value added, Government Office regions, 1995

Region	Percentage
Northern Ireland	34.7
Scotland	34.5
Wales	30.1
South East	29.7
North East	**29.6**
Eastern	28.0
Merseyside	27.0
London	26.4
West Midlands	25.6
UK	25.0
North West	21.7
Yorkshire and Humberside	16.5
East Midlands	16.4
South West	15.6

Source: Government Statistical Service, 1997.

Table 2.9   Proportion of economically active in each region by SOC (%)

	UK	S East	North	Wales	Scotland
Managerial	14.4	16.4	11.3	13.1	12.5
Professional	9.4	11.1	7.6	8.6	9.4
Assoc Professional	8.7	11.3	7.7	8.2	8.0
Clerical	14.2	15.3	12.9	11.6	13.7
Craft	12.5	10.7	14.0	12.9	12.5
Personal/Protective	10.1	9.9	10.4	10.9	10.2
Sales	7.6	7.3	8.5	7.6	7.6
Plant operators	9.4	6.7	11.2	11.1	9.5
Other Occupations	13.8	11.4	16.4	16.0	16.5

Source: NOMIS © Crown Copyright Reserved ONS Statistics (NOMIS), 1996.

3 Defence Closure and Job Loss: The Case of Swan Hunter on Tyneside

JOHN TOMANEY, ANDY PIKE AND JAMES CORNFORD

Introduction

Regional development studies have become increasingly concerned with the economic and social impact of reductions in defence expenditure on dependent local economies (Gripaios and Gripaios, 1994; CEC, 1992). As Nick Hooper and Barbara Butler (1996) recently noted, the "rationalisation" of the defence industries in Britain following the end of the Cold War is estimated to have lead to the loss of some 200,000 jobs in defence contractors and that 'many of these jobs are highly skilled, including R&D and design teams which have in the past been responsible for maintaining and developing a large part of the national technological base'. As they point out, 'Very little is known about what happens to these workers and their skills when defence expenditure falls and jobs are lost' (*ibid.*, p. 149). This chapter tries to provide some answers to the latter question by drawing on a survey of former workers at Swan Hunter's shipyard on the Tyne (see Tomaney *et al.*, 1999). The collapse of the Swan Hunter shipyard in 1993, following its failure to win a key Ministry of Defence order, was a dramatic event which won international attention. This chapter reports the results of a survey of the experiences of former Swan Hunter Workers undertaken in 1995, two years after the receivers were called in.[1]

The Swan Hunter episode

By the end of the 1980s, Swan Hunter was the only remaining active shipyard in the North East of England. At the time of nationalisation, the company, despite a long tradition of merchant shipbuilding, was designated as a warship producer. Given the competitive failure of UK merchant shipbuilding firms and the shrinking size of the sector, remaining producers were increasingly concentrated in the military sector, so that by 1985 – when Swan Hunter was privatised – naval construction accounted for 76 per cent of UK output in the ocean-going vessel segment (Hilditch, 1990).

This segment was relatively sheltered from international competition and naval orders provided a steady stream of work for Swan Hunter through the early 1980s.

The Conservative Government pursued a policy of privatising state-owned shipbuilding firms which were potentially viable in the private sector. For the Government, as well as conforming to its ideological predisposition, privatisation of warship yards also appeared to hold out the possibility of efficiency gains through the creation of a group of competing equipment suppliers to the MoD. It was in this context that managers at the then state-owned Swan Hunter sought to buy the yard from the Government, with the sale completed in 1985. However, from the outset the new company was small and under-capitalised in relation to its main competitors for MoD orders. Despite early successes in winning frigate and fleet auxiliary orders and efforts to modernise the yard through investment in the design capacity, by the early 1990s Swan Hunter faced mounting problems. Efforts to diversify production into the merchant sector failed, reflecting the difficulties experienced by other defence contractors in adjusting to new markets (Hooper and Hartley, 1993). A gap in the forward order book led the company to lay off 1,400 workers in 1992 and the company remained short of a large naval order to sustain its order book.

The largest single piece of equipment on the Ministry of Defence's naval shopping list in the early 1990s was a Landing Platform for Helicopters (LPH) and Swan Hunter was generally regarded as the front runner for this contract. After the Conservative Government called for tenders for the contract, both Swan Hunter and its rival VSEL bid. VSEL, while lacking recent experience in surface vessel production, proposed to collaborate with the merchant shipbuilder Kvaerner Govan of Glasgow.

To universal surprise and amid great publicity, the MoD announced in May 1993 that the order for the LPH would go to VSEL. Within less than two days, Swan Hunter had called in the receivers threatening the remaining 2,700 jobs and, symbolically, the survival of shipbuilding on the Tyne. While unions, politicians and the regional press launched a campaign to save the yard, the receivers looked for a buyer. However, the receivers also began making staff at the yard redundant, a process which continued over two years as the existing frigate contracts were completed. The receivers continued to search for a buyer, while at the same time attempting to retain the design team — regarded as the yard's main asset — on the payroll for as long as possible. The role of the receivers and the positive relationship they formed with the trade unions and workforce was a key factor in allowing the yard to be sold.[2] At one stage the yard was almost sold to the French naval constructor CMN, but was eventually bought in late 1995 by a Dutch concern, THC – the sale being completed only days before equipment in the yard would have been sold off lot by lot.

What happened to the workers?

In May 1995, almost exactly two years after Swan Hunter's went into receivership, a questionnaire was mailed to over 2,200 former employees made redundant, virtually the entire workforce at the time of liquidation.[3] Eventually, 1,645 completed or part completed questionnaires were returned (a response rate of approximately 75 per cent which is high relative to similar surveys; cf. e.g., Gripaios and Gripaios, 1994; Goudie 1996; Noble and Schofield, 1993; Hinde, 1994). Overall, the respondents were broadly representative of the workforce at the time the firm went in receivership although, as in similar surveys, there appears to be some under-representation of unskilled manual workers (Hooper and Butler, 1996).

Workers were made redundant or left Swan's in a number of waves, with the vast majority ending their employment over the two years between the company entering receivership and the survey (see Figure 1). Only one respondent was still employed by the receivers at the site — a security guard. The employment status of the respondents at the time of the survey is summarised in Table 3.1 (broken down by age group) and Table 3.2 (broken down by occupational categories). In addition to standardised and structured questions, respondents were asked to comment in their own words on their experiences since leaving the firm. In the rest of this section we explore the experiences of former Swan Hunter workers both quantitatively and, using the workers' comments, more qualitatively.

The findings of much of the literature on the impact of plant closure in the defence sector and other industries on the workers have been neatly summarised in seven points: 1) many redundant workers cease to be unemployed within a relatively short period, accepting a new job, entering training or education, retiring or voluntarily leaving the workforce, typically within one year; 2) those who remain unemployed are likely to do so for a long time; 3) unskilled and older workers have more difficulty finding work; 4) most redundant workers do not move in search of work; 5) the majority of workers accept lower skilled work and lower pay when starting their new job; 6) the state of the local economy affects the outcome of redundancy; and, 7) the way in which the company conducts its redundancy programme makes a significant difference to the well-being of those made redundant. (Hooper and Butler, 1996, p. 150; see also Goudie, 1996; Gripaios and Gripaois, 1994; Hinde, 1994). The findings of the survey reported here broadly bear out these points. However, we are also able to qualify some of the points and add greater qualitative detail using the workers' own words.

The experience of unemployment

Of the 1,645 respondents, 634 (38.5 per cent) were unemployed at the time of the survey. Of the unemployed group, 422 did not report having had any form of paid work since leaving Swan's, generating a total, from this group alone, of 434 person years of unemployment at the time of survey. Dividing the workers into two roughly equal groups, depending on when they left the firm, we can see that workers who had left the firm by the end of 1993 were slightly less likely to be unemployed than those who left later.

A key factor usually associated with difficulties in finding employment is age. The survey seems to bear these findings out with the highest proportions of unemployed in the 50 to 54 age band – over half of all workers in this group were out of work at the time of the survey compared with an average of 38.5 per cent for the respondents as a whole.

This finding certainly coincides with the experience of many of workers as revealed in their comments. For example, the following quotations give a flavour of many workers' perceptions:

> the new foreign employers in the region seem to have a discriminatory attitude to certain age groups (Former planning engineer with 17 years at Swan's, 47 years old, unemployed).

> After applying for about 40 jobs in various occupations, I feel that my age is a prime factor in companies not being interested enough to employment (Former test manager with 20 years at Swan's, 61 years old, sick).

Even relatively young workers felt that they were at a disadvantage:

> Now at 33 years old I feel I am on the scrapheap and the only jobs I will find will be short-term contracts. (Former electrician with 4 years at Swan's, 33 years old, unemployed).

A second factor associated with unemployment is skill levels. Amongst those respondents classified as unskilled manual workers, over 45 per cent were unemployed at the time of the survey while the rate amongst skilled manual workers was around 42 per cent; rates for managers (18 per cent), clerical workers (29 per cent) and design and technical workers (28 per cent) were much lower, if far from negligible. One frequently voiced perception among the workers, which may help to explain why the differentials, in particular those between skilled and unskilled workers, are not larger, is that that employers did not value skills developed in shipbuilding. For example, one worker wrote that:

In recent interviews the main stumbling block towards full time employment appears to be the fact that my experience has been gained in shipbuilding and I am being pigeon-holed, as though I would be unable to do a job outside shipbuilding (Former process planning technician with 21 years at Swan's, 38 years old, unemployed).

Education and training

A total of 171 of all the respondents had received some retraining and 47 (2.9 per cent) were undertaking training or education at the time of the survey. Further, some 56 of the respondents reported having sought training advice and the majority (39) had found it useful, only few (8) having a negative experience. Some workers were very appreciative of the advice or training that they received:

I found the advice of the Chamber of Commerce very useful and encouraging (Former shipwright with 29 years at Swan's, 53 years old, employed).

Even those for whom the training and advice agencies had not been successful were often appreciative:

I have nothing but praise for job clubs. Very helpful and socially acceptable (Former production/scheduling engineer, 41 years at Swan's, 57 years old, unemployed).

Nevertheless, there seems to be some wariness about the motives for some training providers and the effectiveness of some training provision amongst many of those who received retraining. The following quotes represent common complaints:

Training courses seem to be based on getting people through the qualification rather than what is best for the individual. Training providers, like any other company, are there to make a profit (Former planning technician with 24 years at Swan's, 41 years old, employed).

Have done Government funded courses ..... welding ... decorating ... courses where the unemployed are exploited ... firms getting cheap labour (Former welder, 18 years at Swan's, 36 years old, unemployed).

The biggest swindle is this Training for Work. Who needs to be trained for six months to be a shop assistant? (Former electrical foreman with 30 years at Swan's, 47 year old, employed).

Routes into work

For those in work or self employed, there were a wide range of routes by which work had been found (see Table 3.4 — although the categories are non-exclusive, respondents were asked to identify the *main* route through which they had found work). As has been found in most studies of job search, the largest single group (45 per cent) had found work through word of mouth. More formal mechanisms such as Job Centres, newspaper adverts and employment agencies were a significant route into work and job clubs seem to have been useful to some. The single second largest category was "other" mainly comprised of speculative letters and visits.

Falling out of the labour market: the sick and retired

Only 15 of the former Swan's workers who responded to the survey (or less than 1 per cent) had left the labour market through retirement (unsurprisingly these were mainly older workers near the normal age for retirement). By far the larger group to have left the labour market were the sick with around 12 per cent of respondents (197) reporting being sick at the time of the survey, again with a clear bias towards older workers (for example, almost half of all the 60-64 year olds reported being unable to work through sickness and were claiming sickness benefit). Some directly linked their sickness to the stress of being made redundant: "after leaving Swan Hunters I have become very stressful; and as a result have had a stroke" (Former turner with 18 years at Swan's, 52 years old, sick).

In and out of work: insecurity and instability in the North East's flexible labour market

Out of 634 respondents who were unemployed at the time of the survey, some 212 had had some work in the period since leaving Swan Hunter. Age is, again, a significant factor here, with unemployed respondents over 40 years of age at the time of the survey less likely to have worked since leaving Swan's than those 40 or under — only 30 per cent of the over 40s having worked, compared with 43 per cent of those 40 or under. This is born out in many of the workers' responses. For example,

> Done Jobsearch, done Jobclub. Made full use of facilities provided, i.e. phone, letters, CVs to no avail, and (found that) age limits exist (Former plumber, 55 years of age at time of survey, unemployed).

Perhaps surprisingly, however, skill levels seem to have had little impact on whether those unemployed at the time of the survey had worked since leaving Swan's: the managerial category stands out with 81 per cent of the unemployed having worked, but this is followed by the unskilled manual workers (72 per cent) and the remaining categories bunched together with around two thirds of unemployed workers having worked since leaving Swan's.

The majority of the 212 of those unemployed at the time of the survey who had some experience of work most had had only one job (122) or two jobs (56) in the intervening period, usually lasting a few weeks. One worker, a former Coppersmith, however, had nine separate periods of employment between leaving the firm at the end of November 1993 and filling in the questionnaire in May 1995. The average job length for those that were unemployed at the time of the survey was a less than 16 weeks (even for those in work at the time of the survey it was only 26 weeks).

Even where former Swan's workers were able to find paid employment, this was often in the form of temporary contracts. Of those 683 respondents that were working for an employer at the time of the survey, 294 (or 43 per cent) were on temporary contracts. The perceptions of many of the former workers was that short term contracts were becoming the norm. For example, one respondent wrote:

> Seven jobs since leaving Swans, all short-term contracts, there's no security, you can never plan ahead (Former GMBATU member with 18 years at Swan's, 36 years old, employed).

Experiences of temporary employment certainly left some workers bitter:

> Employers just use you to suit their purposes, and discard you before you can claim any benefits such as pension rights (Former skilled machinist with 33 years at Swan's, 49 years old, unemployed).

Another wrote:

> As all jobs are temporary it is hard to make any long-term plans (Former welder with 11 years at Swan's, 30 years old, employed).

Some workers felt that such short term contracts were not beneficial to their employers either:

> Short term contracting is no good for either employer or employee – no commitment to company on the one hand and none by the company on the other (Former deputy project weapons manager with 3 years experience at Swan's, aged 45 at the time of the survey).

Many of those who were in employment contrasted the labour practices of their new employers with those they had experienced at Swan's:

> (In new job, compared to Swan's) everything is monitored — efficiency, scrap, downtime, absence, etc. etc. Headcount adjusted downwards if sales fall ... completely different environment, no-one is 'carried'! (former works engineer with 2 years at Swan Hunter, aged 28 at the time of the survey).

> I now work in the retail sector, namely dealership (motor). Working practices are very different, "Do it or else" being prevalent .... the only other opportunities I could find with a decent wage would have been outside the area (Former Designer with 11 years experience at Swan's, aged 30).

> I'm working twice as many hours for less money, but what's the alternative to life on the dole? It's nae picnic! (Former Pipeworker with 15 years at Swan's, aged 46, employed).

The lucky ones? permanent employment and self-employment

The proportion of each broad skill grouping in permanent employment varies widely from 68 per cent for the managerial group, 58 per cent for the clerical group and 56 per cent for the design group but as low as 47 per cent for the skilled manual workers and just 50 per cent for the unskilled manual workers. Age made a smaller difference with the ratio of temporary to permanent contracts slightly higher for those over 40 compared with those 40 years old or less.

Working away

One response to the closure of the shipyard, as in other examples of plant closure, was to travel in search of work. Of 718 respondents who were in employment or self employed at the time of the survey, 563 were still in the North East, 109 were elsewhere in the UK and 32 were working abroad (14 gave no indication). Those working abroad included 4 managers, 16 design or technical workers and 25 skilled manual workers and one clerical worker. Working abroad was most common among the design and technical workers with over 12 per cent of respondents in that category that were working doing so abroad. These workers were scattered across a range of locations including France, Belgium Singapore, Saudi Arabia, Sri Lanka, the Falkland Islands and Australia. The largest groups however, were in the Netherlands (mainly Rotterdam), and in the United States (predominantly in San Diego, California).

A larger group was working in the UK but outside the North East. Once again the locations were dispersed with significant concentrations in Glasgow (25), Southampton (13), Barrow in Furness (12), Rosyth (7) and London (7). A further 15 workers did not indicate a location or wrote that they were working "all over" the UK. In this group too, there is a clear skills bias with over half of all the managers travelling outside the region and over a quarter of the design and technical workers compared with just ten per cent of skilled manual workers and even lower proportions of unskilled manual and clerical workers (see Table 3.5).

The financial benefits of travelling to find work seem clear (see Table 3.6). Taking those in employment who gave information on both their Swan's salaries and their current salaries, those who stayed in the North East saw their average weekly take home pay fall from £242.47 at Swans to £221.75 in their new job. Meanwhile those who left the region saw a rise in average weekly take home pay from £246.64 to £390.73. These gains, however, have to be set against the costs of travelling to find work. The experience of working away from the North East would appear to be stressful for many of the former Swan's workers:

> Any more contracts abroad, and I'll be divorced (Former Section Leader with 11 years at Swans, aged 48, employed in the Netherlands).

> Having left a steady job with paid sick and holidays and getting home every night, I have now become a Technical Gypsy living away from home ... (Former section leader design engineer (electrical) with 35 years at Swan's, 51 years old, employed at Rosyth in Scotland).

> I leave home at 3am and travel to work for 5 and half hours on a Monday morning — stay in a B&B and return Friday evenings. Still, I have a job (Former senior designer with 26 years at Swan's, 43 years old, employed).

> Although the salary is better, it does not reflect the loss of company pension, life insurance, private health insurance, etc. .... It also involves a lot of travelling (Former project quality assurance manager with 26 years at Swan's, 51 years old, employed in Barrow-in-Furness).

The self-employed

A total of 35 of the former workers were self-employed at the time of the survey. A wide range of occupations were found although the largest group (seven) were self-employed electricians; other occupations included taxi driver, publican, video salesman and financial advisor. The self employed were slightly more likely to be working outside the region, but this probably reflects a slightly higher skill profile for this group rather than any association between entrepreneurial activity and "getting on your bike". Once again, age is a factor for the self-employed, indeed for older workers the sense that employers are unlikely to hire them may be the driving force leading to self employment:

> Nobody wants to employ old wrinklies like me full-time. I have had to become a "TINA" consultant (TINA = There Is No Alternative)   (Former Q.A. Manager with 14 years at Swan's, 58 years old, self-employed).

Another worker makes the same point more explicitly, but with a less happy eventual outcome:

> Because of my age, I thought the best chance of employment was to start a business. This went very well for several months, until a major investor decided that the risk was too great and walked away (Former manager sheet metal/joiners shops, 30 years at Swan's, 57 years of age, sick).

Re-using Swan's skills in the labour market

Several respondents reported problems with having the skills that they had developed at Swan's accepted in the wider labour market. The following quotations give a flavour:

> Too many companies seem to want/prefer graduates with paper qualifications but little or no experience — presumably they can be 'moulded' to the desired shape with little resistance (Former subcontracts section leader with 15 years at Swan's, 46 years old, employed).

> Prospective employers appear to be ignorant of the high-tech nature of shipbuilding and tend to stereotype shipyard workers making it difficult to break into other fields (Former project inspection manager with 18 years at Swan's, 46 years old, employed).

When those workers who were employed or self-employed at the time of the survey were asked directly whether they were using the skills that they had acquired at Swan's, only 432 out of 718 (60 per cent) replied positively. However, many of the 40 per cent who were not using their skills felt bitter about this. For example, one design worker wrote:

I would not have normally left engineering for sales but as I only had this opportunity I had no option but to accept (Former Draughtsman with 8 years at Swan's, 27 years old, employed).

Conclusions

The closure of Swan Hunter represented a major blow for the local area. Although many of the workers did find new jobs in the period between being made redundant and our survey, these were less secure and (if local) less well paid. Others were less lucky and remained unemployed. It would be wrong to suggest that there were no workers who felt that they were better off at the time of the survey compared to when they worked at Swan's. Some respondents told happier stories, for example:

> After being made redundant I spent nearly 12 months out of work, but since gaining employment through my training course, I really enjoy my work. (Former welder with 21 years at Swan's, aged 39, employed).

Nevertheless the overwhelming sense of the personal experience of the workers is negative. When asked directly if they felt better off, the same or worse off overall at the time of the survey than when they were working at Swan's, the majority of those who expressed a view in every category indicated that they felt worse off, unsurprisingly led by the group that were unemployed, followed by the sick, those in training or education, and those with employment (see Table 3.7). Yet even in this latter group of "lucky ones", over half felt that they were worse off than when they were working at Swan's.

On top of this human tragedy, the closure also saw the break-up of the collective technical expertise of the Swan Hunter design team. While the majority of these skilled workers were able to find new jobs, this was often in the context of long-distance commuting or work abroad. The longer term impact of such a brutal 'deconstruction' of a workplace on the UK's technology base remains unclear. It appears, however, hardly likely to be positive.

Notes

[1] The survey was funded BBC Radio 5 Live and undertaken with assistance of the Confederation of Shipbuilding and Engineering Unions. We are especially grateful to Eddie Darke, former secretary of the Swan Hunter CSEU for his support.

[2] According to Derek Horsfield of the receivers Price Waterhouse: 'One or two of our managers have got so enthused they'd be happy to stay there for ever'. His colleague Ed James added: 'One of team's gone native — he wears a hard hat a lot' (quoted in Tighe 1993, 7). Despite a growing mutual respect, the yard's unions successfully sued the receivers for unfair dismissal of the workforce, in the process establishing a precedent in employment law relating to insolvency and redundancy.

[3] This large sample was made possible by the co-operation of the Swan Hunter trade unions.

References

CEC (1992) *(The economic and social impact of reductions in defence spending and military forces on the regions of the Community)*, CEC: Luxembourg.

Goudie, I. (1996) 'Redundant Defence Workers Survey', Report commissioned by the Strathclyde Defence Industries Working Group, March 1996.

Gripaios, P. and Gripaios G. R. (1994) The Impact of Defence Cuts: The Case of Redundancy in Plymouth. *Geography* 79, 32-41.

Hilditch, S. (1990) Defence Procurement and Employment: The Case of UK Shipbuilding. *Cambridge Journal of Economics* 14, 483-496.

Hinde, K. (1994) Labour Market Experiences Following Plant Closures: The Case of Sunderland's Shipyard Workers. *Regional Studies* 28, 713-724.

Hooper, N. and Butler, B. (1996) A Case Study of Redundant Defence Workers. *Defence and Peace Economics* 7, 149-167.

Hooper, N. and Hartley, K. (1993) UK Defence Contractors. Adjusting to Change. Research Monograph Series 3, Centre for Defence Economics, University of York.

Noble, M. and Schofield, A. (1993) After Redundancy. In The Factory and the City: the Story of the Cowley Automobile Workers in Oxford, edited by T. Hayter and D. Harvey. London: Mansell, 231-255.

Tighe, C. (1993) 'Receivers hold out hope for Swans', *Financial Times*, 13th November, p.7.

Tomaney, J., Pike, A. and Cornford, J. (1999) Plant closure and the local economy: the case of Swan Hunter on Tyneside. *Regional Studies*, 33, 401-412.

Table 3.1   Former Swan Hunter Workers by Age Group and Employment Status (number and per cent of age group)

Age Group	Unemployed		Retired		Sick		Training and Education		Employed or Self Employed		Not Stated		Total
	No.	%	No.	%	No.	%	No.	%	No.	%	No.	%	No.
20-24	3	27.27	0	-	0	-	0	-	8	72.73	0	-	11
25-29	26	27.96	0	-	0	-	4	4.30	61	65.59	2	2.15	93
30-24	58	31.18	0	-	0	-	8	4.30	118	63.44	2	1.08	186
35-39	58	29.15	0	-	6	3.02	4	2.01	129	64.82	2	1.01	199
40-44	75	36.23	0	-	8	3.86	7	3.38	110	53.14	7	3.38	207
45-49	105	38.75	1	0.37	18	6.64	9	3.32	130	47.97	8	2.95	271
50-54	113	45.38	0	-	43	17.27	7	2.81	80	32.13	6	2.41	249
55-59	118	50.86	1	0.43	50	21.55	5	2.16	54	23.28	4	1.72	232
60-64	64	44.44	2	1.39	62	43.06	3	2.08	11	7.64	2	1.39	144
65-69	4	19.05	10	47.62	7	33.33	0	-	0	-	0	-	21
Not Stated	10	31.25	1	3.13	3	9.38	0	-	17	53.13	1	3.13	32
Total	634	38.54	15	0.91	197	11.98	47	2.86	718	43.65	34	2.07	1645

Table 3.2 Former Swan Hunter Workers by Skill Band and Employment Status (number and per cent of skill band)

Skill Band	Unemployed		Retired		Sick		Training and Education		Employed or Self Employed		Total
	No.	%	No.	%	No.	%	No.	%	No.	%	No.
Managers	12	18.46	-	-	7	10.77	1	1.54	44	67.69	65
Design and Technical	56	28.00	-	-	10	5.00	9	4.50	121	60.50	200
Skilled Manual	492	41.59	14	1.18	154	13.02	31	2.62	465	39.31	1,183
Clerical	16	26.67	-	-	4	6.67	2	3.33	38	63.33	60
Unskilled Manual	47	45.19	-	-	19	18.27	4	3.85	32	30.77	104
Total	623	38.65	14	0.87	194	12.03	47	2.92	700	43.42	1,612

Table 3.3 Employment status at the time of the survey by date of leaving Swan's

Year left Swan's:	Not Stated		Unemployed		Retired		Sick		Education and Training		Employed or Self Employed		Total
	No.	%	No.	%	No.	%	No.	%	No.	%	No.	%	No.
1994 or later	14	1.77	354	44.81	2	0.25	60	7.59	16	2.03	344	43.54	790
1993 or earlier	18	2.14	275	32.66	13	1.54	135	16.03	30	3.56	371	44.06	842
Not Stated	2	15.38	5	38.46	0	-	2	15.38	1	7.69	3	23.08	13
Total	34	2.07	634	38.54	15	0.91	197	11.98	47	2.86	718	43.65	1645

Table 3.4 How were jobs found

Main source of Job	Number of Workers	Per cent of all Workers Employed
Job Centre	84	11.70
Agency	62	8.64
Word of Mouth	322	44.85
Newspaper	84	11.70
Job Club	11	1.53
Other (mainly speculative letters and visits)	137	19.08
Not stated	18	2.51
All workers Employed	718	100.00

Table 3.5 Skill band and location for workers employed or in self employment at the time of the survey

Skill Band	Not UK		Rest of UK		North East		Total
	No.	%	No.	%	No.	%	No.
Management	4	9.09	23	52.27	17	38.64	44
Design and Technical	16	12.70	35	27.78	75	59.52	126
Skilled Manual	25	5.23	48	10.04	405	84.73	478
Clerical	1	2.50	1	2.50	38	95.00	40
Unskilled Manual	0	-	2	6.67	28	93.33	30
Total	46	6.41	109	15.18	563	78.41	718

Table 3.6 Earnings Compared with Swan's Earnings by Location at the time of the Survey (number of respondents who gave both figures and per centages of location groups)

	Location					
	Not North East		North East		Total	
Current Earnings are:	No.	%	No.	%	No.	%
Less than at Swan's	28	22.95	312	67.97	340
More than at Swan's	94	77.05	147	32.03	241
Total	122	100.00	459	100.00	581

Table 3.7 **Perceptions of personal situation at the time of the survey compared with when working at Swan's**

	Not Stated		Better		Same		Worse		Total
	No.	%	No.	%	No.	%	No.	%	No.
Not Stated	11	32.35	2	5.88	3	8.82	18	52.94	34
Unemployed	44	6.94	7	1.10	20	3.15	563	88.80	634
Retired	9	60.00			1	6.67	5	33.33	15
Sick	37	18.78	3	1.52	6	3.05	151	76.65	197
Education and Training	3	6.38	8	17.02	5	10.64	31	65.96	47
Employed or Self Employed	12	1.67	147	20.47	168	23.40	391	54.46	718
Totals	116	7.05	158	9.60	203	12.34	1159	70.46	1645

Figure 3.1

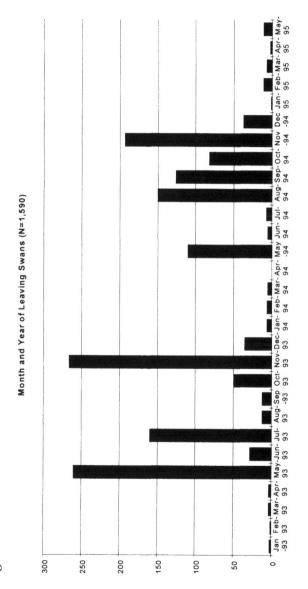

Source: CURDS Survey.

4 Working in the Business Family: Micro-business Livelihoods in the North East and the South East of England

SUSAN BAINES AND JANE WHEELOCK

Introduction

Enterprise and entrepreneurship are part of 'our new ambition for Britain', declared the Chancellor of the Exchequer, Gordon Brown, in his budget speech of March 1998. 'In the new economy', he explained, 'jobs will come not simply from having a small number of large businesses but a large number of small and growing businesses' (Financial Times, 18th March 1998). North eastern counties have consistently produced relatively low levels of business start-ups but the number of new businesses in the region has recently hit a record high, according to a report by the Northern District Society of Chartered Accountants. The northern president of that organisation observed a new "feel good factor among companies – partly because Prime Minister Tony Blair has described up and coming companies as the bedrock of a successful enterprise economy".[1] Nearly twenty years after the Thatcher administration began to promote the benefits of an 'enterprise culture', small businesses are still expected to be dynamic, flexible and, most of all, the creators of new jobs.

Around a third of employment in the EU is now estimated to be in micro-businesses (employing fewer than ten) and such very small firms therefore determine the working conditions of millions of women and men across Europe (Scase, 1995; Gray, 1998). Yet although there has been overwhelming interest in the *numbers* of jobs attributable to the formation and growth of new small firms, there has been comparative silence about their actual work and employment practices. Such enquiry, we contend, is long overdue if we are to contribute to deeper understanding of the extent to which indigenous growth strategies are feasible – and, very importantly, at what costs to the individuals involved.

In this chapter we draw upon survey data and in-depth interviews to disclose three aspects of the livelihoods of small business owners, their families and employees. First, we highlight owners' experiences of employing which sit uneasily with confident expectations of job creation. We add to evidence that levels of business owners' interest in employment growth are generally low and show that there is movement in both directions between the employing and non employing categories. We also highlight the extremely personal nature of employment relationships for micro-business owners. Second, we show how intimately small business owners' capacity to avoid the worst pitfalls of becoming employers is associated with the complex of network relationships in which they are embedded. Business owners in that minority who actively seek employment growth are those who network most enthusiastically with other businesses and organisations. Third, and perhaps most controversially, the significance of the family for providing labour to ..small firms is emphasised. Family work can be a vital resource without which a struggling business would fail to survive but there are costs as well as rewards for businesses, individuals and families. Gender divisions of labour are, typically, extremely traditional.

An assessment of the relative merits and problems of markets, hierarchies (including those in families) and networks is drawn upon to illuminate the 'whether to employ / how to employ' issue that is assessed in this empirical work. Family and non family employees may be alternatives but they are not direct equivalents (Pollack, 1985). Self exploitation within the business family and sacrifices made by some individuals, especially women participating in business alongside their husbands, are, we conclude, much under reported aspects of micro-business life.

Small business employment: myth and reality

A quarter of a century ago small businesses were regarded as part of an out-moded traditional economy and studies of working lives understandably overlooked them. It has been observed that the seminal report of the Committee of Inquiry on Small Firms (the Bolton Report) published in 1971 devoted only four out of more than 400 pages to employment (Rainnie, 1989; Stanworth and Gray, 1991). How best to assess the numbers of new jobs that may be attributed to the small firm sector has become the subject of detailed and sometimes acrimonious debate since then (Robson and Gallagher, 1993; Hart and Hanvey, 1995; Robson, 1996).

Small firms are seen as generators of jobs and wealth and, while their employment practices are little discussed, they are usually assumed to be unproblematic as Rainnie (1992) noted. A rare dissenting voice from the policy arena has been raised by the Trades Union Congress (1997). Businesses employing ten or fewer, they concede, have come to account for a higher proportion of jobs in the UK than in the past. Unfortunately jobs in micro-businesses, they argue, are likely to be of poor quality and short duration. Moreover, much of the widely heralded rise in the numbers of small firms is actually accounted for by self-employed individuals who are not employers at all.

The TUC's last point is consistent with government figures which show that, in the 1980s, the 'decade of enterprise', it was the self-employed without employees whose numbers rose most dramatically (Campbell and Daly 1992). Hakim found among new recruits to self-employment evidence of "hitherto untapped interest in and potential for entrepreneurship" (1989, p.289) but that was certainly not the whole story. In parts of North East England hardest hit by unemployment there was a rise in self-employment in marginal service sectors (Storey and Strange, 1992). People entering self-employment "strove to establish working lives in the cracks of a local economy devastated by mass unemployment" (MacDonald, 1996, p.432). New small firms, from this perspective, look much less dynamic and exciting. They have come to prominence as large corporations delegated risk further down the line, to subcontractors, to smaller firms, to a more flexible workforce and to the self-employed. The overall result has been a transfer of business risk to the household (Wheelock and Mariussen, 1997).

The un-stated assumption in the TUC report is that the non-employing self-employed will inevitably remain non-employing. More typically, it is assumed that at least some of the non employing self-employed are potential or embryonic employers. Very little is actually known about the processes by which self-employed people without employees become employers, or not. From such evidence as does exist it seems likely that dependence upon families is common. Meager (1992) observed that self-employed and business owning women and men in the UK are more likely to be married than the working population as a whole. He speculated that this could be explained, in part at least, by the importance of business support from a spouse. A pilot study conducted by Wheelock in Wearside in the North East of England found that work for the small business unit and for the family unit were so closely inter-related as to justify the use of the term 'familial economic unit' (Wheelock, 1992).

Surveys published since then have indicated that participation in small firms by owners' families is the norm rather than the exception in the UK generally (Rosa *et al.* 1994; Poutziouris and Chittenden 1996). The idea of the 'familial economic unit' has been supported by Ram and Holliday (1993) who argue from ethnographic research that the family is crucial to understanding the pattern of social relations within small firms but that family can be both a resource and a constraint. The picture accepted in popular imagination is that family labour is typical of Asian businesses but when Jones *et al.* investigated this in a study of Asian, white and Afro-Caribbean business owners they found that family labour was a traditional and important feature of small businesses which did not stem from ethnic origin so much as from the intrinsic disadvantage of small business enterprise (Jones *et al.* 1994).

Employing family and non-family: a transaction cost approach

Analysis in terms of the transaction costs incurred in markets as opposed to the governance structure of firms or families provides a useful starting point for the issues this chapter addresses. It was Oliver Williamson (1975; 1987) who developed Ronald Coase's (1937) insights as to how the existence of firms as economic institutions can be explained in a perfectly competitive market. There are costs involved in using markets, costs which may be reduced by utilising hierarchies within firms (or families) instead. Such transaction costs are generated on the one hand by the 'bounded rationality' of agents operating in the market. On the other hand, markets provide the possibility for opportunism on the part of participants. (In the case studies discussed below, for example, there was an instance of criminal behaviour by one individual newly recruited by a young business). The transaction cost approach argues that it may be cheaper to do business within a hierarchical organisation such as a firm, where transaction specific investments in obtaining knowledge can be preserved and where the pursuit by economic actors of their own advantage can be better handled. It is costly to write and enforce contracts to cover even some of the contingencies that may arise in a market exchange. A transaction cost approach indicates why the 'visible hand of management' (Chandler, 1977) may supplement the invisible hand of the market.

Transaction cost analysis is able to comprehend, as more classical economic analyses cannot, a real world in which economic actors may be devious (Granovetter, 1985). Nevertheless, according to Granovetter, the approach as normally applied is limited because it overlooks 'the role of concrete personal relationships and structures (or "networks") of such relationships in generating trust and discouraging malfeasance" (p.490). Empirically, he argues, 'there is evidence all around us of the extent to which business relations are mixed up with social ones' (*ibid.*, p.195). Powell too suggests limitations in the transaction cost approach. He notes that the philosophy within networks is quite different from that of markets – where the strategy is to drive the hardest possible bargain – and from hierarchies – where action is shaped by concerns with career mobility. 'In networks, the preferred option is often one of creating indebtedness and reliance over the long haul' (Powell, 1991, p.270).

Robert Pollak (1985), extends the transaction cost approach to the family, arguing that the governance structure of the family holds a number of advantages over both the market and non-family firms, which all 'flow from its ability to integrate .... with pre-existing, ongoing, significant personal relationships' (p.585). Indeed, it is noticeable that many non-family firms rely heavily on family metaphors to smooth relations with employees (Roberts and Holroyd, 1992; Holliday, 1995). In terms of incentives, the family commands rewards and sanctions which are not open to other institutions, and there are likewise monitoring advantages because key behaviours such as diligence and work habits are more likely to be observable. Altruism within the family limits opportunistic behaviour, and loyalty cements relations advantageously. What Pollack fails to discuss, however, is the significance of gender hierarchy within families.

An empirical study of micro-businesses

The empirical study upon which we report in this chapter investigated family involvement in the survival, maintenance and growth of business service businesses[2] in two contrasting urban locations: the city of Newcastle upon Tyne in the North East of England and the new town of Milton Keynes in the South East. These locations were chosen for contrasting socio-economic profiles in the 1980s, the period in which most of the businesses we studied were founded. Out of 280 UK towns and cities Milton Keynes has been ranked third on a range of prosperity indicators. Newcastle upon Tyne, was ranked a poor 210th (Champion and Green, 1992).

A telephone survey conducted with 200 micro-businesses (100 in each location) formed the starting point of our investigation into family involvement.[3] Respondents were asked about co-ownership and employment. While these do not, as we will discuss later, by any means exhaust the possible ways in which family members participate together in business, they are relatively unambiguous and readily ascertained in a short telephone enquiry. From the information so collected it is possible to delineate the broad contours of formal family involvement in micro-businesses in those sectors and locations.

Businesses were not included unless they had been in existence for at least two years but no other age criteria were imposed. The 200 business service micro businesses turned out to be overwhelmingly young businesses. Their median age was only 6.0 years in 1994 and fewer than one in ten had existed before 1980. Because of the young age of the businesses it not surprising that family businesses which had been in families for more than one generation were hardly represented. Only two were owned by more than one member of a family of different generations.

The investigators' expectation, based on the Wearside work, was that familial involvement would be most characteristic of the less advantaged location. This was not the case. Numbers of businesses formally involving husband and wife in some way (either as co-owners or employees) proved to be extremely similar across locations, at 30 per cent in Milton Keynes and 33 per cent in Newcastle. Businesses formally involving *any* family member were also similarly represented in each location (39 per cent in Milton Keynes and 41 per cent in Newcastle). Thus 40 per cent of the micro-businesses in both locations were family businesses in the sense that they formally involved family members as co-owners or employees. By far the most common family relationship in business was husband and wife. The overview from this data is that family participation, as co-owners or employees, was practised by a minority, but a very substantial one, of micro-businesses in business services. This was almost the same in locations with very different socio economic characteristics.

The results we draw upon for the analysis in the rest of this chapter are based on the more detailed and context sensitive information obtained from the two stages of face to face interviews which were conducted with subsets of the 200 telephone respondents. The first of these stages was an interviewer administered questionnaire with 104 owner-managers drawn at random from the larger sample. Detailed information on divisions of labour and sources of help and support in business and domestic life was collected. Although the questionnaire was mainly in multiple choice format, open ended questions were also asked about attitudes to business growth and to support networks. The final stage consisted of in-depth interviews with owner-managers (and where possible another significant person in the business) of 34 of the businesses already interviewed. People were invited to identify critical incidents in the recent life of their family and their business and to talk in detail through those incidents they considered most significant (Chell, 1998).

The remainder of this chapter is divided into four sections. First, we report patterns of family and non-family work and employment and business owners' intentions towards further employment growth. Second, we examine business networking and show that active networking was most practised by that minority of business owners with a positive attitude to employment growth. These data are based on the extended questionnaire conducted with 104 owners. Then, we turn to case study work to address more difficult and complex questions around the process of becoming and remaining employers for owner managers. Families and networks are shown to be crucial to this process because they can ensure, through long-term relationships, loyalty, trust and reliable information. This tends to confirm both Pollack's presumption that it may be advantageous for businesses to call on family and Powell's emphasis on networks with their normative, not legal, forms of sanction. Finally, in the concluding section, this material is brought together to suggest an understanding of relationships within small businesses as a 'moral economy'. When it is the household that manages business organisation risk we see something which may seem incongruous in a developed economy approaching the end of the millennium. That is the reinstatement of aspects of old ways of working, particularly dependence on the family and highly gender stereotypical divisions of labour.

Work and employment in micro-businesses

Only a quarter of the 104 micro-business owners who participated in the second stage of our research reported that they had employed anyone at start-up. Three fifths were employers at the time of the interview. However, there was by no means a straight-forward record of progress from non employing to employing. Of those businesses which did not employ at the time of interview, more than a quarter had employed at least one person in the past. We will return to this theme later when we examine the experiences of employing recounted to us in in-depth interviews. We now continue to examine the questionnaire data. We first report the employment intentions and recruitment practices of the employing micro-businesses and then turn to ways in which human resources were called upon by means other than direct employment.

The answers to a series of open ended questions about plans and opportunities for the businesses in the immediate future were carefully studied and sifted. Respondents' intentions to grow were assessed from this information. Just a quarter (25 per cent) indicated that they were set on course for some employment growth. This is the group we denote by the shorthand term 'growth enthusiastic'. The largest group by a long way (39 per cent) were rejecters of growth who resolutely did not intend to grow at all in the future, although they may have done so in the past. A further 29 per cent indicated that they may seek employment growth in the longer term. Finally, 8 per cent intended to grow with non-standard paid labour rather than direct employment.

Forty respondents (representing 46 per cent of those with domestic partners) reported what we describe as 'high' involvement of that partner in the business. These 40 out of 104 businesses were husband and wife businesses in the sense of *either* being co-owned by husband and wife (21) *or* employing the owners' spouse (13) *or* having very substantial unpaid daily practical input from a spouse (6). Daily unpaid work on the part of a spouse was quite rare but when the cases reporting some form of unpaid work for the business by an owner's spouse are added to these cases of 'high' spouse involvement, three quarters of the businesses in which the respondent lived with a spouse may be described as, in some way, husband and wife businesses.

The division of business tasks between husbands and wives was overwhelmingly traditional. With very few exceptions women who participated in business alongside their husbands, whether as co-owners, employees or unpaid helpers, performed the support and service roles associated with women in the labour force as a whole. There was no difference in this respect between the northern and the southern location.

Of the 40 non-employing businesses, 26 (65 per cent) reported that they used freelance or casual labour at least occasionally. However, the use of such non-standard workers was not associated exclusively with the smallest businesses. On the contrary, the practice was slightly more common (70 per cent) among the employing businesses. This may seem counter-intuitive but it accords with the findings of Hakim (1989) who notes that casual labour is a supplement to, not a direct substitute for, direct employment.

In summary, family involvement in micro-businesses was most commonly that of spouses. Instances of the participation of some kind by spouses in the micro-businesses we interviewed were many. The pervasiveness of spouse participation in the businesses tends to confirm Pollack's predictions. The evidence further suggests a characteristic of family input which was not part of Pollack's analysis. That is the presence of a gender hierarchy. While spouses were called upon as sources of labour, that was not the only way in which flexible labour could be marshalled by the owner of a micro-business. Use of *paid* informal labour was common practice. Family were quite widely employed but networks activated for recruitment purposes were based on business, not family, contacts. (For a fuller report on these findings, see Baines and Wheelock, 1997).

Networking and employment growth

We sought to understand, from the face to face interviews, to what extent business owners drew upon, for business purposes, advice, support or assistance from people who were not part of the business or the family. Some form of contact with local institutions was quite high. For example, nearly 40 per cent reported some contact with their local Chamber of Commerce in the past three years. Yet there was a marked lack of enthusiasm for formal and institutional networks even from people who nominally participated in them. Comments such as "we are in the Chamber of Commerce but we wonder why", were far from unusual. That, however, was not the whole story. Loosely linked groups, often composed of other owners of small businesses and of colleagues known through former employment, were highly valued. These were variously esteemed for practical advice relevant to business development and for more intangible but strongly desired moral support. "I get the social contact I would normally get in a large organisation", declared one marketing consultant, male partner in a husband and wife team. "Most of my friends are small business owners. We meet once a week and talk business issues like the in-house team of large organisation", noted a male sole trader in the design business.

Overall, rather more than a third (37.5 per cent) of the businesses were either extensive or moderate participants in some kind of network which was either business related or a combination of business and social.[4] In other words a substantial minority of the business owners in the locations and sectors we studied recognised and acted upon a need for business expertise, practical support and friendship from people outside the business and the family. Presence in the 'growth enthusiastic' group had been tabulated against 'high spouse involvement' (Table 4.1) and 'networking activity' (Table 4.2).

Businesses in the 'growth enthusiastic' group were less likely than other businesses to have reported that spouse involvement was high (either co-ownership, formal employment or daily unpaid practical support). The difference is not significant at the $P = 0.05$ level. There is, however, a statistically significant association between the practice of networking outside the family and presence in the 'growth enthusiastic' group. That minority of businesses which were set on course for growth were, it seems, distinguished by their owners' readiness to seek out resources beyond the family.

Families, hierarchies, networks and micro-business job creation

The questionnaire data indicated that, overall, women who participated in businesses alongside their husbands typically played a supporting role to the husband's professional or craft skills. The in-depth interviews reveal example after example of women who adapted to the changing needs of businesses, sometimes sacrificing individual careers to do so. Twelve out of the 34 critical incident interviews included some account of a difficult decision by husband and wife around how time should be allocated between work for a business, child care and earning wages outside the business. Most typically, in eight out of these twelve cases, the decision involved some increase in the wife's support for the business. Her contribution could be work for the business (three cases), income earned outside to support the business (one case), or both work and income (four cases). In another case a woman had recently taken a full-time job outside the business in order to catch up with her interrupted career. Her earnings as a manager in the public sector had made the business founded by her husband sustainable in its early years and she had then worked for it full-time in an unusually egalitarian partnership with him. The other three cases concerned men's contribution to their business and family. One husband who had been made redundant considered, but decided against, caring for his child full-time in order to enable his wife to concentrate on her business. "He's an old fashioned sort of guy", she explained. There were two cases in which women's and men's actions somewhat bucked this traditional pattern. In both cases a husband with a young family decided to reduce his commitment to his business in the interests of his wife's separate career.

In order to better understand the conflicting pressures upon marital partnerships, businesses and individuals, let us look at the words of just two of the women who worked alongside their husbands, performing vital supporting tasks. Eric and Pat were co-owners who suffered the loss of their major client soon after they made a substantial outlay in new computer equipment. Pat had once regarded her role as co-owner as largely nominal. However, she took on all the office work for the business because Eric could no longer afford an employee. She hated office work. "I'd rather walk the streets in the pouring rain," she declared. In fact, she did that too as she had a part-time job selling household goods door to door to supplement the family income. She also cared for their four young children and felt resentfully that she was 'like a single parent' because of Eric's long hours.

Even more overt conflict between a women's own desires and the needs of a micro-business is seen in the case of Bella. This is how she explained that she left her 'lovely little job' as a saleswoman in a department store to work full time for the typesetting business set-up by her husband Russell:

> It got to that stage where he was ringing me. I used to work sort of ten till three – he was ringing me at quarter to three and saying "can you come in". So I was leaving there at 3 o'clock and coming in here. Erm, and it just, I think it was worse than a full-time job. It was like two. There was either, stop working here and employ somebody to do my job, or stay at [the department store]. And I was only working part-time, so that really wasn't feasible.

Despite her comment that employing somebody else while she continued to work part-time elsewhere was not feasible for the business, Bella also expressed some doubts about working alongside her husband. He often, she said, 'gets uptight' and would take his anger out on her and she hated the idea of her teenage sons seeing their father behave in this way. His outbursts did not, however, spill over into life at home. If they ever did, she would not continue to work for the business. Particularly clear in her case, then, are dangers of family conflict as well as advantages of family support, both of which Pollack points too. These themes recur throughout the interviews with various levels of intensity.

Whereas Wheelock (1992) found in her earlier pilot study that *family* flexibly contributed to business survival, the Newcastle and Milton Keynes data indicate that such flexibility is largely, but not exclusively, spouse based. The commitment and resilience of women participating in businesses with their husbands cannot be overstated. Yet, the 'family business' of husband and wife is a hierarchy of a very particular kind in which gender relations are overwhelmingly reproduced in traditional fashion. The demands of business life put pressure on the institutions of gender (Wheelock and Mariussen, 1997). Characteristically, it is traditional solutions which are reached within households with greater flexibility on the part of women.

The bald figures cited suggest very considerable turbulence in owner managers' employing experiences. This turbulence is dramatised in the critical incident interviews. Altogether 19 out the 34 respondents chose to describe employment related situations. Four had positive experiences to relate and eleven, in contrast, described difficult and sometimes painful situations. A further four related to us both positive and negative employing experiences.

Only a few studies have focused on the nature of employment and employment related problems for small firms (Curran and Stanworth, 1981;

Scott *et al.*, 1990; Rainnie, 1989; Goss, 1991; Atkinson and Meager, 1994). Of these only Atkinson and Meager directly address the initial employment of others by the self-employed. This is the 'entry level' in their model of four stages of engagement with the labour market by small firms. The entry level, they report, is characterised by extreme informality and by the employment of spouses, other family and known contacts. It often follows a long period of 'making do' with casual workers, buying in services and working long hours. The in-depth interviews with owners of micro-businesses indicate that, while this account is indeed an accurate overview of the behaviour of the self-employed who do begin to employ, it is rather bland. It glosses over the very complex and often painful process through which the business woman, man or partnership struggles.

Business owners are often unable to make a fully rational choice about taking on an employee because they have limited knowledge of the capacities and skills of prospective employees. In other words there is, in the parlance of transaction cost analysis, a problem of 'bounded rationality' for agents operating in the market. It is to obviate this problem that owners of small businesses take on people they know well, including members of their families (Scott *et al.*, 1990). That this was common practice for service sector micro-businesses in Milton Keynes and Newcastle was borne out in the data cited above. In the critical incident interviews, there were five cases where family members other than spouses were, or had been, directly employed and another in which the employment of a family member was planned for. Some of the mixed benefits of a family employee are clear from Sally's words about her step-daughter:

> She's more interested in the fact that Mike forgot to send her a Valentine card! But I need her because of the flexibility. On a Wednesday, because I'm on site, she spends Wednesday morning here. She transfers the calls home and she answers the calls all afternoon. And she only puts a time sheet in for the morning. She says 'no it's the least I can do', you know. So, you know, there's got to be flexibility.

In such cases family members were valued as employees because they could be trusted and because they worked from obligation and may not even demand fully rewarding for all their work. Examples of this kind are unsurprising and fit the scenario outlined by Pollack and the patterns found in the small body of empirical literature already referred to. The accounts given to us also include cases in which bringing into a business family members, or other well known individuals, had a harsher outcome.

The most spectacularly unsuccessful venture into employing family was the experience of husband and wife business owners Eric and Pat already mentioned. They had employed Eric's brother, Mark, who had happened to be in need of a job when an employee left. As Eric pointed out, this avoided the need to go through the expense and trouble of advertising, sifting applications and interviewing prospective employees. Mark, however, found the day to day work menial and below his capabilities and he resented taking instructions from his brother and sister-in-law. The friction caused a family rift which had never been healed. Eric and Pat dismissed Mark when they lost their major client. It was not the dismissal itself but the whole of what Pat called on several occasions 'the Mark thing' which hurt Pat and Eric so deeply. "I lived it slept it, breathed it" said Pat of this period of bitter conflict.

Tony and Isobel similarly employed people who were well known to them. They had worked alongside Tony in his previous job. The owners became so dissatisfied with the work of one of these employees that they eventually dismissed her. The other employee left a few weeks later with many recriminations and bitter words. Here Isobel and Tony talk of their relief at no longer being employers:

> Isobel
> It's just so wonderful just to concentrate on the work and we work together very very well, and we don't have to spend all that time trying to brief people and make them understand the objectives and basically there's so much time investment in people and not just time, it's just [long pause] being just the two of us we can just be ourselves as well and just get on with things and....
> Tony
> Not be on show [pause] and I know it's quite difficult to explain.

The experiences described by Tony and Isobel (who are an extreme but by no means isolated case) are not easy to reconcile with most small firms studies, where the emphasis has tended to be on the quantitative potential of new small enterprises to become employers. It is true that there are a few examples in the large quantitative literature of research which acknowledges the uneven employment providing potential of small firms and that many owner managers neither seek nor desire growth (Storey, 1994; Gray, 1998). Contrary to policy makers' expectations, Storey and others found from analyses of more than 600 firms' records at Companies House that no direct link existed between taxable profits and job creation (Storey *et al.*, 1987). The painful inter personal conflicts which may underlie some owner managers adversity to growth are, nevertheless, almost entirely invisible.

As we saw from the survey data, business contacts were the most popular personal source of recruitment information. The in-depth interviews suggest, moreover, that the employers who experienced the most harmonious working relationships with their employees recruited neither total strangers from advertisements nor individuals known as friends or colleagues. They chose people about whom high quality information has been carefully gathered through industry networks. Two young and growing businesses where this was the case were run by Dominic, and by the husband and wife team Bella and Russell, already referred to. Bella and Russell decided to recruit a sales representative to develop new business while Bella concentrated on the clientele she had already built up. Their search was long and anxious as they had no way of ascertaining the trustworthiness and competence of people sent by the Job Centre. Bella described the frustration of lacking reliable information on which to make such an important decision.

> You know they might have come with a nice list and said well [I have] this client and that client, and all the rest of it. And you didn't know them from Adam. It was a big risk.

Then Russell heard through his contacts in the industry of a woman with just the right experience and attitude.

> He just knew she would work and she does work.

Dominic similarly explained how he recruited an employee whose competence was such that it was becoming possible to achieve some organisational specialisation which would enable him to develop the business.

> I knew what sort of calibre and personality we'd get.

Dominic found a worker he could trust only after an unpleasant venture into employing when his first employee disappeared with the business's vehicle and later turned out to have been involved in criminal activities. That was an unusually dramatic story, but a very serious message which is conveyed again and again in the experiences reported to us is that becoming an employer involves unfamiliar ways of relating to other people and new owner managers are often ill prepared for this in the extreme. Good quality information about prospective employees is vital but not easy to obtain and the consequences, when a mistake is made, can be severe.

Employment was repeatedly described as a serious commitment and something not to be undertaken lightly. Micro-business owners insisted that they did not want to 'hire and fire' although they sometimes thought that others behaved in that way. As one employer of eight expressed it, "I think if we were bigger and sort of harder people we might take on lots more people and weed them out and that sort of thing". This sense of personal responsibility sometimes extended even to people working for micro-businesses on a casual and short term basis. Bernard, for example, told us how he and his wife had rejected the use of direct employment after they had been cheated by employees and a non-family business partner. They were confident, however, that they could successfully restructure their enterprise, working together from home with highly trusted individuals in an 'arms length' relationship with the business. "What we will do is actually develop something, organically, a cottage industry, but with people that we know and we trust, and we'll do it via technology." In another instance, Roland ran a translation service in which the work of ninety eight casual translators was co-ordinated from a home/office. Some of the casual workers were very regularly called upon and relationships, on occasion, became similar to more formal employment. Roland reported that the loss of a major customer left one of the casual translators, who was a former colleague, without work upon which he depended. Roland made an effort to target more work in this individual's area of expertise because he felt a sense of obligation.

Powell (1991) points out that just as economists come to view firms as governance structures, firms themselves seem to be blurring their boundaries with a whole range of quasi market arrangements like collaborative ventures, subcontracting, and so on.... Post-Fordist management practices in large firms have led to the introduction of a whole array of 'flexible' organisational arrangements, which, as some see it, have 'hollowed out' the traditional Fordist corporation (Jessop, 1994). The cases of Bernard and Roland indicate that even micro-businesses themselves may participate in flexible forms of production which blur the boundaries of the organisation. Such innovative, tele-mediated ways of working with a dispersed workforce were, however, not typical of the micro-businesses in this study. Much more characteristically, their flexibility involved a blurring of boundaries between business and family which we have called the reinvention of the 'traditional' business family.[5] From this perspective flexibility, that holy grail of modernising economies, actually often involves the reproduction within modernity of seemingly pre-modern practices in household organisation and gender divisions of labour.

Conclusions

Family participation is a component of small business life in the UK which is only now becoming widely recognised. Our evidence shows that family involvement in business service businesses is most usually between husbands and wives and it is more extensive than the most easily measurable factors of ownership and employment suggest. The business family may be understood as a special kind of hierarchy characterised by altruism and loyalty. These advantages come over in cases where owners emphasise the reliability of family members who cheerfully adjust their daily lives around the unpredictable workload characteristic of very small businesses and sometimes do not even demand full payment for all their work. But as Pollak notes, there are disadvantages associated with each of these areas. One disadvantage is that there can be a lack of appropriate skills within the family. Our survey data showed that growing businesses were most likely to network outside the family, suggesting some recognition of the weakness of relying on family alone. Another, more disturbing, disadvantage is that conflict can spill over from one sphere to another, where family instability may become a source of weakness. Painful consequences are dramatised in the most extreme case reported above where the employment of an owner's brother led to an irreparable family rift.

This chapter has offered new insight, based on empirical research, into some aspects of work, recruitment and employment in small businesses. Examination of markets, hierarchies (including families) and networks has helped to understand their behaviour against a background of economic restructuring. Becoming an employer for the first time involves unfamiliar ways of relating to other people and this can be so difficult for some new owners of small businesses that they withdraw from employing altogether. For those who do employ, as has long been known of agricultural workers, the relationship between employer and employee in any micro-business is inevitably a highly personal one. Trust and loyalty are key components in such a relationship: families and networks can be used to help to ensure these.

E.P. Thompson used the term 'moral economy' to characterise the highly personal, symbolic and hierarchical flavour of medieval markets, and extended its use to an understanding of the household (Thompson, 1971). The search to maintain levels of profitability that has gathered force in the advanced capitalist nations since the 1973 oil crisis signalled the end of the golden era of post-war economic growth, and has led to a ground swell shift from hierarchy to markets within the giant organisations of state and corporation. Micro-businesses and the self-employed have grown both as a response to downsizing, as larger organisations have shed labour, and in response to the opportunities that the 'hollowing out' of organisational hierarchies can bring. Whether the result of positive or negative forces, owners find themselves uniquely exposed to the full blast of market forces. It is scarcely surprising some should attempt to protect themselves with the reconstruction of a form of the moral economy.

The family can, as the Wearside pilot study predicted, be a vital resource for micro-businesses (Wheelock, 1992). This is true in the North East and in a location with very different characteristics. Sometimes, to an observer, it appears that women in many micro-businesses experience considerable exploitation, often working hard at tasks for which they have little taste and, in some instances, sacrificing their own careers in the interests of the family project of the business. These observations were made in the 1990s. Moreover, they were made in a growing business sector not normally associated with tradition nor, indeed, with the family. No matter how fervently politicians wish to see small firms as new, dynamic and innovative, the uncomfortable evidence is that very old ways of working characterise the livelihoods they provide.

Notes

[1] The Journal, 28/11/97.

[2] Business services covered were three groups: a) knowledge intensive professional services such as design and consultancy   b) office services e.g. secretarial, bookkeeping and recruitment and   c) craft which consisted mainly of printing and commercial photography.   We excluded professional services with institutionally enforced barriers to entry such as solicitors, accountants and architects.

[3] A population list of businesses in the selected business services employing between 0 and 9 employees in both locations was created. In Newcastle this was compiled from data acquired by telephoning 400 businesses from *Yellow Pages*, the Tyne and Wear Economic Intelligence Unit data base, the Chamber of Commerce Directory, press advertisements and some professional directories.   In Milton Keynes the well researched *Commerce Directory* was used.

[4] 'Moderate' participation denotes some active engagement and does not include, for example, people whose only network involvement was nominal membership of a Chamber of Commerce.

[5] For further development of this argument and a comparison between modernised family businesses in Norway and their reinvention in Britain, see Mariussen *et al.*, 1998.

References

Atkinson, J. and Meager, N. (1994) 'Running to stand still: the small firm and the labour market', in J. Atkinson and D. Storey (eds). *Employment, the Small Firm and the Labour Market,* London: Routledge, 28 - 102.

Baines, S. and Wheelock, J. (1997) *A Business in the Family: an Investigation of the Contribution of Family to Small Business Survival, Maintenance and Growth,* Leeds: The Institute of Small Business Affairs, Research Series Monograph No. 3.

Ben-Porath, Y. (1980) 'The F-connection: families, friends and firms and the organization of exchange', *Population and Development Review,* 6, 1-30.

Campbell, M. and Daly, M. (1992) 'Self-employment into the 1990s', *Employment Gazette, June,* 269 - 292.

Chandler, A.D (1977) *The Visible Hand: the Managerial Revolution in American Business,* Cambridge: Harvard University Press.

Chell, E. (1998) 'Critical incident technique', in G. Symon and C. Cassell (eds)., *Qualitative Methods in Organisational Research: a Practical Guide,* London: Sage, 51 - 72.

Champion, A. G. and Green, A. E. (1992) 'Local economic performance in Britain during the late 1980s: the results of the third Booming Towns study', *Environment and Planning A,* 24, 2, 243 - 272.

Coase, R. (1937) *The Firm, the Market and the Law,* Chicago: University of Chicago Press.

Curran, J. and Stanworth, J. (1981) 'Size of workplace and attitudes to industrial relations in the printing and electronics industries', *British Journal of Industrial Relations,* 19, 1, 14 - 25.

Goss, D. (1991) *Small Business and Society,* London: Routledge.

Granovetter, M. (1985) 'Economic action and social structure: the problem of embeddedness', *American Journal of Sociology,* 91, 3, 481 - 509.

Gray, C. (1998) *Enterprise and culture,* London: Routledge.

Hakim, C. (1989) 'New recruits to self-employment in the 1980s', *Employment Gazette, June,* 286 - 297.

Hart, M. and Hanvey, E. (1995) 'Job generation and new and small firms - some evidence from the late 1980s', *Small Business Economics,* 7, 2, 97-109.

Holliday, R. (1995) *Investigating Small Firms: Nice Work?,* London: Routledge.

Jessop, B. (1994) 'The transition to post-Fordism and the Schumpeterian workfare state', in R. Burrows and B. Loader (eds). *Towards a Post-Fordist Welfare State?,* London: Routledge, 14 - 37.

Jones, T., McEvoy, D. and Barrett, G. (1994) 'Labour intensive practices in the ethnic minority firm', in J. Atkinson and D. Storey (eds). *Employment, the Small Firm and the Labour Market,* London: Routledge, 172 - 205.

MacDonald, R. (1996) 'Welfare dependency, the enterprise culture and self-employed survival', *Work Employment And Society,* 10, 3, 431-447.

Mariussen, Å., Wheelock, J and Baines, S., (1997) 'The family business tradition in Britain and Norway: modernisation and reinvention?', *International Studies of Management and Organization,* 27, 3, 64 - 85.

Meager, N. (1992) 'The characteristics of the self employed' , in P. Leighton and A. Felstead, *The New Entrepreneurs,* London: Kogan Page, 69 - 99.

Pollak, R. A. (1985) 'A transaction cost approach to families and households', *Journal of Economic Literature,* 23, 2, 581 - 608.

Poutziouris, P. and Chittenden, F. (1996) *Family Businesses or Business Families?*, Leeds: Institute for Small Business Affairs in association with National Westminster Bank.

Powell, W.W. (1991) 'Neither market nor hierarchy: network forms of organisation', in G. Thompson, J. Frances, R. Levacic and J Mitchell, *Markets, Hierarchies and Networks*, Sage: London, 265 - 276.

Rainnie, A. (1989) *Industrial Relations in Small Firms: Small Isn't Beautiful*, London: Routledge.

Rainnie, A. (1992) 'Flexibility and small firms', in P. Leighton and A. Felstead *The New Entrepreneurs,* London: Kogan Page, 217 - 236.

Ram, M. and Holliday, R. (1993) 'Relative merits: family culture and kinship in small firms', *Sociology, 27, 4,* 629 - 648.

Roberts, I. and Holroyd, G. (1992) 'Small firms and family forms', in N. Gilbert, R. Burrows and A. Pollert (eds). *Fordism and Flexibility: Divisions and Change,* London: Macmillan, 154 - 169.

Robson, G. and Gallagher, C. (1993) 'The job creation effects of small and large firm interaction', *International Small Business Journal,* 12, 1, 23 - 37.

Robson, G. B. (1996) 'Unravelling the facts about job generation', *Small Business Economics, 8, 5,* 409-417.

Rosa, P., Hamilton, D., Carter, S. and Burns, H. (1994) 'The impact of gender on small business management: preliminary findings of a British study', *International Small Business Journal, 12, 3,* 25 - 32.

Scase, R. (1995) 'Employment relations in small firms', in P. Edwards (ed). *Industrial Relations: Theory and Practice in Britain,* Oxford: Blackwell, 569 - 595.

Scott, M., Roberts, I., Holroyd, G. and Sawbridge, D. (1990) 'Management and Industrial Relations in Small Firms' Sheffield, Department of Employment.

Stanworth, J. and Gray, C. (eds). (1991) *Bolton 20 Years on: The small firm in the 1990s*, London: Paul Chapman.

Storey, D. (1994) *Understanding the Small Business Sector*, London: Routledge.

Storey, D., Keasey, K., Watson, R. and Wynarczyc, P. (1987) *The Performance of Small Firms*, London: Croom Helm.

Storey, D. and Strange, A. (1992) *Entrepreneurship in Cleveland 1979 - 1989: A Study of the Effects of the Enterprise Culture*, Sheffield: Department of Employment.

Thompson, E. P. (1971) 'The moral economy and the English crowd in the eighteenth century', *Past and Present,* 50, February, 76 - 136.

Trades Union Congress (1997) *The Small Firm Myths: A TUC Analysis of the Small Firm Sector within the UK Economy*, Economic and Social Affairs Department, TUC.

Wheelock, J. (1992) 'The flexibility of small business family work strategies', in K. Caley, F. Chittenden, E. Chell and C. Mason (eds). *Small Enterprise Development: Policy and Practice in Action,* London: Paul Chapman, 151-165.

Wheelock, J. and Mariussen, Å. (1997) *Households, Work and Economic Change: a Comparative Institutional Perspective*, Boston: Kluwer Academic Press.

Williamson, O. (1975) *Markets and Hierarchies, Analysis and Antitrust Implications: a Study in the Economics of Internal Organization*, New York: Free Press.

Williamson, O. (1987) *Antitrust Economics: Mergers, Contracting, and Strategic Behaviour*, New York: Blackwell.

Table 4.1 High spouse involvement by enthusiasm for growth

Spouse Involvement	Growth enthusiastic	Other growth intentions	All businesses
	n = 26	n = 78	n = 104
High	7 (26.9%)	33 (42.3%)	40 (38.5%)
Moderate, low or absent	19 (73.1%)	45 (57.7%)	64 (61.5%)

Table 4.2 Network participation by enthusiasm for growth

Network participation	Growth enthusiastic	Other growth intentions	All businesses
	n = 26	n = 78	n = 104
Extensive or moderate	5 (57.7%)	24 (30.8%)	39 (37.5%)
Limited or none	11 (42.3%)	54 (69.2%)	65 (62.5%)

significant at P = 0.5

PART II
POLITICS AND CULTURE IN TRANSITION

5 Region-Building in the North East: Regional Identity and Regionalist Politics

CHRIS LANIGAN

Introduction

The North East of England has been the site of a number of political developments over the last 13 years or so that could be described as regionalism. There are two, arguably distinct, aspects this regionalism. First, there has been institution- or capacity-building. This type of regionalism has attempted to create structures of political and economic governance designed to enable political and economic elites in the region to respond to particular public policy priorities, particularly those connected to economic development. The Northern Development Company (NDC), North of England Assembly of Local Authorities (NEA), North East Chamber of Commerce and Northern Business Forum (NBF) are some examples of such institutions. Such institutions are 'regionalist' in the sense of their reliance on the idea of the region (whether that be the North East or the North East plus Cumbria), as a unit where there are cross class and cross locality common interests. However, their emergence does not indicate either a decentralisation of political or economic power in the United Kingdom, or indeed a wish to see such decentralisation.

However, in the 1990s a second type of regionalism, an explicitly politically decentralist one, has emerged. It is this type of regionalism that this chapter will focus on. This movement for a devolution of political power, and for regional government, has its most visible expression in the Campaign for a Northern Assembly but has also, to a certain extent, been taken up by other actors such as the NEA, some MPs, unions and so on.

This chapter looks at the relationship between this type of regionalism and regional identity. It suggests that while regional culture and identity may be factors in promoting both types of regionalism other factors such as political, institutional and economic change at the national and European level are important, as is the perceived self interest of particular groups and factions. However, while identity may not be the fundamental trigger of regionalism, it is a resource that regionalists of both types can attempt to use to legitimise and promote their aims.

We assume here that regional identity is a socially constructed category, open to conscious attempts to define and re-define it. We shall pay particular attention to two such attempts to build and define a regional identity. In recent decades there have been two particular coherent, elite led 'identity projects' which have sought to encourage regional identity to evolve in particular directions.

Regional culture and regional identity

The last twenty years have seen major changes in cultural practices in the North East. As one generation has passed and a new one arrived the social base of the population has inevitably changed. Such change has been accelerated and exaggerated by the huge changes in work practices such as the collapse of the old extractive and manufacturing industries, and some of the work based collective culture that went with them, and the continued expansion of female participation in paid work. There has also been the effect of increased exposure to national and global culture through television, film, magazines, computers and music. Some traditional cultural practices already in decline at the time of Durham University's survey of regional culture in 1973 (Townsend and Taylor, 1974; 1975) such as whippet racing, leek growing and miners' galas have no doubt continued to decline but other arguably "regional" cultural traits such as accent/dialect and attachment to heavy drinking have evolved rather than declined.

However, we are not so much concerned with cultural practices as they exist but with what the people of the region think are the cultural practices that define regional identity. In the imagined community (Anderson, 1983) or structure of feeling (Williams, 1961) of the North East, what are the shared regional memories, activities, and ways of thinking? What is it that people think makes the North East the North East and the people North Easterners?

This chapter is going to examine two different, and arguably conflicting ongoing 'top-down' attempts to encourage particular conceptions of regional identity in an attempt to promote particular economic and political goals. On the one hand, we have an attempt being made by both public and private sector senior figures involved in economic regeneration work to promote the idea of a vibrant, entrepreneurial, successful region. We will call this the *market oriented identity project*. On the other hand, we can discern a competing tendency for some regional intellectuals and politicians to promote the idea of the region as inherently opposed to what are seen as 'Southern' 'Thatcherite' values due to morally advanced collectivist and socialist values deeply ingrained in the regional psyche. This view tends to lead to support of the political project of regional government or devolution and we will call it the *regional community oriented identity project*. We will now examine these in detail.

The market oriented identity project

This regional identity project has focused on the economic regeneration needs of the region. The North East's position of relative deprivation in terms of the UK as a whole dates back to the 1920s. By the 1970s it was being suggested that aspects of regional culture were barriers to economic modernisation and recovery. The pit villages and pits, and towns based around one major industry such as ship building, engineering, or steel, were seen to have bred a population antithetical to entrepreneurialism and insufficiently geographically or socially mobile for a successful modern economy to take root (Northern Region Strategy Team, 1977). With the collapse of much of the declining industrial base of the region from 1977 onwards, the failure of North Easterners to create their own jobs through entrepreneurialism – a failure contrasted with high rates of self employment in the South – seemed to suggest to some that cultural change had to be encouraged.

This industrial collapse coincided with a growth in the volume and sources of potential 'inward investment' – in other words, multinational firms seeking to set up production facilities outside their home market. Inward investment provided an obvious partial solution to the unemployment crisis in the region but how could inward investors be attracted to come to the region? The answer was though place marketing. As Sue Wilkinson's (1992) study *Towards a New City?* shows, the perceived need to compete with other locations to attract inward investment led to attempts to change both internal regional identity and the external perception of the region. A need was seen to project an image of a place where foreign companies could do business, where they would find happy, contented workers and where (very importantly), their senior managers would be happy to move to. Certainly, aspects of the region's image as seen from outside – depressing pit heaps, limited cultural life, poor educational levels, were seen as barriers to inward investment (and also to tourism).

Within the North East the influence of this place marketing campaign was felt in several ways: slogans, flagship developments, events, billboards/adverts, and through the media. The slogans of the Tyne and Wear Development Corporation, 'Building the New North East' and the

NDC 'The Great North' sum up the ideological thrust of this identity project – namely that the North East was being (successfully) renewed and that this process would be completed by the hard work and honest toil of the region's people. This renewal was implicitly and explicitly linked to a particular reading of the 19th century history of the region. As the critic, Hudson (1991, p.47) explains:

> The idea of the resurgence of the 'Great North' conjures up images of a past golden age and a revival of the positive aspects of the region's position in the nineteenth century as a centre of economic growth and technological progress.

The market oriented project was able to attempt to legitimise itself through trying show that the NDC/Development Corporation strategies were not alien, Thatcherite impositions but fully in keeping with regional culture, history, myth and symbols.

But this was also a project to change identity. For outsiders, the aim was to promote the image of a recovering, industrious region, full of attractions such as a hard working, well disciplined workforce; copious sites of high culture (Theatre Royal, Northern Sinfonia; and high quality consumption sites (Durham City Centre, Metro Centre, Teesside Retail Park, the Bridges etc), set in beautiful countryside (Tynedale, Northumberland Coast, North York Moors) with inexpensive, quality housing and good road, rail and air communications. This would attract jobs from inward investors and spending from tourists.

For regional inhabitants, the aims of this market oriented identity project were arguably more subtle. There was certainly the long unrealised aim of 'modernising' old attachments to collectivist lifestyles and ushering in a new region where self employment, business success, innovation and risk taking were admired and accepted, not suspected. If the region's people could be persuaded (by using the history of 19th century entrepreneurship plus evidence of successful regeneration in the present) that "real" jobs didn't have to involve bashing bits of metal together for a large employer but could be self employed service sector, telesales, or computing jobs, then, the leaders of this strategy would argue, economic recovery in the region would be much advanced.

There was also an attempt to eliminate regional pessimism that saw continued decline as inevitable, and suggested that the region could do nothing to help itself, apart from the Jarrow March strategy of petitioning London. The market-oriented identity project, with its attempts to link up with the histories of 19th century enrepreneurialism could point to a time when people from within the region achieved things for themselves without looking to London or other outsiders to do things for them. With levels of regional aid reduced and Thatcher having told many of the region's industrial great and good to their faces at Durham Castle in 1985 that they were a "bunch of wingers" and that there would be no government bail outs of floundering industries, this attempt to promote a more self-reliant, self confident form of regional identity had some logic to it.

Wilkinson's study concentrates on Newcastle but she looks at one genuine regional body, the Northern Development Company and one sub-regional one (the Tyne and Wear Development Corporation). There are, however, other examples which show that this identity project was being attempted throughout the North East. Durham County Council never loses an opportunity to draw a veil over its coal mining past with its twin Prince Bishop heritage/tourism/roadsigns campaign and the First Class County sport theme (promoted though the annual cross country athletics event as well as the cricket team). On Teesside, the Teesside Development Corporation went as far as physically turning the River Tees from a muddy estuary into a freshwater lake with white water canoe slalom to combat Stockton's image 'problems'. The Hartlepool marina development can be seen in the same way – erasing a previous image of industrial dereliction in favour of one that will (in theory) attract both tourist and inward investors alike.

Limitations

Perhaps the main limitation of this identity project is that the image it put forward was:

> not ... a straightforward description of reality but ... part of a promotional advertising campaign intended to bring about some of the changes which it claims have already happened (Hudson, 1991, p.48).

This inevitably leads to occasions where continuing problems of unemployment and poverty undermine the credibility of the regeneration strategy leaders as agents who can define the regional identity of the North East.

Even at the political and ideological height of this project, the late 1980s, well informed critics such as Dr Fred Robinson at CURDS and then Durham University still appeared in the media highlighting the gap between rhetoric and reality. The fragility of the project was shown by the hysterical criticism that Robinson and those like him received. Upwardly spiralling crime figures and the 1991 Tyneside Riots, however, could not be swept under the carpet or blamed on 'politically motivated academics'. The power of this identity project to change people's sense of the North East's regional identity suffered as a result.

This project was also limited by aspects of resistance to the cultural engineering it was attempting to introduce. Top down identity projects work best at a sub-conscious level and among those relatively open to socialisation such as the young. Heavy handed, explicit attempts to reform regional identity are likely to be seen by many as a threat to their identities. Wilkinson (1991, pp.182-185) notes the resistance to attempts to re-invent the traditional 'little waster' cartoon figure Andy Capp, for example. There is no doubt also that this identity project has not changed the mindset of all of its intended recipients: the Panorama programme on Shildon a couple of years ago which found teenage boys still expecting to work in jobs like their fathers even though such jobs had disappeared with the closure of the town's rail wagon works a decade before.

Political implications

Several academic commentators were in no doubt that the market-oriented identity project was explicitly politically Thatcherite as well as economically so. Hudson, for example, claims that the North East's status as a regional economy dominated by previously nationalised, heavily unionised industries meant that if the region could be portrayed as a success story, it would prove that Thatcherite economic policies could work anywhere. (Hudson 1991, p.46). Taylor (1993, p.144) hints that the market-oriented identity project may have been intended to increase Conservative electoral support in the region, but failed to do so.

The strongest link between Thatcherite economic aims, political aims and the associated identity project is made by Byrne (1989; 1992) who suggests that the attempt to promote individualist readings of regional identity was a deliberate attempt to undermine the collectivist, working class, and pro-Labour identity of the region's people in the hope of thus breaking the Labour Movement – both party and unions as a result.

However, the results of policies are not always as intended and while the "Great North" identity programme may be a "unionist" one (Taylor, 1993, p.144) telling a people to cast off their dependence culture, harness their energies and return to Victorian industrial prosperity by their own efforts may lead to people seeking to reduce their dependence on the political centre in London, which they come to see as being out of touch with their needs and holding back their progress. This was arguably a consequence of Thatcherite "self-reliance" rhetoric when applied to Scotland and it may have boosted similar thinking in the North East generally, even in some (minority) circles in the business community. As long ago as 1972, when early elements of the market-oriented identity project were being put into place it was suggested that the diffusion of higher aspirations might be a source of social conflict. For it is fairly well established that, rather than merely being a cause of low achievement, the lowering of aspirations provides a means whereby those lowest in the reward structure may accommodate their expectations with their achievement; if these expectations and aspirations were to outrun the possible means of realising them, then a widespread 'regional relative deprivation' may ensue. One does not spell out in detail the forms that this might take (a Northern Separatist Movement), for the sociological implications of [the economic development elite's] claims have obviously escaped the notice of those employing the developmental framework. (Rowntree Research Unit, 1972).

In the long term, the market-oriented identity project may therefore contribute to discourses in North East society in favour maximum political self-reliance.

The regional community identity project

This alternative regional identity project is one that opposes most aspects of the market oriented identity project, and positions itself on the political left. We will term it the *Regional Community Identity Project* because it promotes the idea that there is a mass-level regional identity, which sharply differentiates the North East from other parts of England, and particularly from the dominant Southern centred statist, monarchical and middle class oriented "English" national identity (see Taylor, 1993). This identity project is also community oriented because it is hostile, or at best ambivalent about the effects of the free market on culture and community life. Existing culture is seen as a source of pride (though capable of criticism and improvement in parts), not of weakness. Much stress is put on collective values and remembering the past radical, socialist history of communities as a source of strength against encroaching individualism / Thatcherism.

The mouthpieces of this identity project are not marketing agencies and redevelopment quangos but academics, authors, and playwrights, an early example of this identity project being articulated is the book, *Geordies: Roots of Regionalism* (Colls and Lancaster, 1992). In *Geordies*, academics and playwrights attempt to thresh out what the region is culturally, politically and geographically in order to point to an alternative future to Thatcherism, that alternative gaining its legitimacy from its basis in regional identity.

As Colls and Lancaster (1992, p.ix) put it,

we are simply trying to gather up the strands of what we have been in the past in order to realise more fully what we can be in the future .... The future from present perspectives looks bleak. The North East's human and material resources have been squandered because it is invited to share an identity [Thatcherite Englishness] which imagines the real nation lives somewhere else. We have to reclaim our resources in order to govern ourselves properly and appropriately. Those who can remember, complain about the loss of 'community spirit'. They talk as if this is inevitable, something to do with 'modern life.' *But it doesn't have to be ....* As with 'community', the North East has tended to believe that there is a dreadful inevitability about its economic decline. It is as if this is a price that Geordies must pay: the former success created pride and pride must be punished. One more we have to say that this erosion of pride has been a disgraceful squandering of resources. *Again, it doesn't have to be like this.*

Most of the early exponents of this identity project locate regional identity in working class led experience, the pit village, ship yard and engineering works. As Williamson (1992, p.155) puts it:

What unites the various strands of the broader regional pattern is a richly textured sense of a hard - worked past, whose marks evoke not only bitterness and poverty, but greatness, too, of industrial achievement and the pride that went with it.

Turning to drama, we can see the same themes strongly represented. The explicit role of the regional community identity project as a device to legitimise left wing politics on the basis of their compatibility with the region's own history and the alien, southern origin of Thatcherism is a recurring theme. In Plater's *Shooting the Legend* (1996) the young, individualistic joyrider rejects the idea of regional culture as a laughable, passe joke until the his elders and betters extend the love of the regional, collectivist community to him. The joyrider dredges up the words of a local folk song from his memory and so realises that the regional collective memory is in him (as in the audience too who are invited to sing along). The message is clearly that the collective regional memory can resist Thatcherite individualism, providing the basis for a civilised, community based future. Amber Film's *The Scar* (1997) embarks on a similar course. The collectivist memories of the mining villages are invoked, particularly community solidarity of the 1984-1985 strike, and though both mother and son are tempted by the seductive representative of Thatcherite individualism (who couldn't even lay down his work to attend his mother's death bed) in the end, the values of community and family prevail.

Again, in the Tyneside Mystery Plays (1997), the Cycle opens with the ghost of T. Dan Smith confronting Sir John Hall at the Metro Centre. Hall suggests that history simply gets in the way of 'progress' but Smith insists that "without a memory we do not know who we are, and who we can be." The rest of the cycle illustrates various points in Tyneside history and present, again making the link between historical memory, identity and future politics.

However, a notable trend which can be seen in some of the writing in Northern Review (e.g. Colls, 1995; Lancaster, 1996), in the *Tyneside Mystery Plays*, and in the *Treasures of the Lost Kingdom of Northumbria* exhibition (1996) is a broadening of the definition of regional culture. In addition to dialect and accent and the working class civil society of the pit, shipyard, Labour Party and union meeting, the 'Lit and Phil' endeavours of the 19th century bourgeoisie, the Northumbrian pipes, and the Dark Ages Northumbrian Kingdom (with its bishops, monasteries, learning and illustrated gospels) have been admitted to the regional memory / identity.

Limitations

The regional community identity project is vulnerable to criticism on a number of grounds. The first is that seeking to define a regional identity is an inherently flawed project. Post-modernists deride such attempts as exercises of power which create "meta-narratives" and create a group of excluded people in opposition to those included. Some on left would also see this identity project as obscuring class divisions within the region, and reject that idea that under "internal colonialism" almost the entire regional population is part of the exploited class.

A second set of criticisms which have rather more threat to the popular appeal of this identity project are those which see it as based on excessive levels of nostalgia and on an over-romanticised view of the working class. If we take Mark Hudson's account of his year in Horden, *Coming Back Brockens,* for example, not only the current social problems in the East Durham coalfield are chronicled but some of the myths about the past are debunked as well. For example, Horden had always had a 'nice end' and a 'nasty end', and the culture of self-learning and union activity had always been those of a minority. There are the criticisms of the maleness of this regional identity (something the project's founders try hard to deal with but perhaps not with full success). Also, there seems to be an inability yet to fully deal with the new working class of the North East, the army of telesales people and the office workers, and build regional myths around them, let alone the middle classes in their rural and suburban hideaways. The nostalgia element of this identity project worries some who would see the region as at the cutting edge of culture and industry. *The Crack*'s review of Plater's *Shooting the Legend*, for example, applauded the play but said that the crowd singalong section was cringeworthy and embarrassing, while a senior development professional bemoaned the role of *Our Friends in the North* in bolstering old stereotypes.

A further criticism is the one that the originators of the Regional Community Identity Project are no more qualified to tell the people of the North East what their identity is than are the development and marketing professionals behind the Market Oriented Identity Project. This argument suggests that those behind the project are engaged in personal searches for the roots which they have lost through social and geographic mobility, and their prescriptions are no less self-interested than the development professionals. Indeed some participants in organisations that defend aspects of 'traditional' culture in the North East see both identity projects as equal, politically inspired threats.

Politics

The political baggage of the Regional Community Identity Project are fairly clear. As we have already suggested, it is strongly anti Thatcherite. However, it is not necessarily pro-Labour in every respect. The non-participative, pro planning strain of Labourism, which broke up communities in the 1960s and 1970s is clearly at odds with this identity project, as is the side of Blairism pitched at the Daily Mail reading, Home Counties, middle class. This project is pro-regional government, but perhaps sees building and raising levels of regional consciousness as its role in achieving that.

Regional identity and regional government

The last few years have seen the idea of regional government move up the political agenda in North East England. This may have been inevitable given the advent of the concept of the Europe of the Regions and the debate on devolution for Scotland and Wales. However, the formation of the Campaign for a Northern Assembly (CNA) in 1992 and their subsequent activities, along with interest shown by the local authorities regional group, the North of England Assembly of Local Authorities (NEA), has propelled the idea of a devolved regional government for the North East / Northern Region to a higher position on the political agenda than elsewhere in England.

Interestingly, the pro-regional government forces in the region have made surprisingly little use of arguments based on regional identity. The NEA's documents, in particular, base their arguments on the principles of accountability, efficiency, and policy effectiveness. Interview based research with some of the NEA's leadership suggests there are a number of reasons for this. Perhaps the most significant reason is that the NEA's leadership are tied into the market oriented identity project through their membership of the board of the NDC. In any case, as council leaders, the NEAs leadership are likely to value "solid" achievements in falls in unemployment more than "airy-fairy" cultural considerations and so if trying to change the image of their area is necessary to bring jobs (as they see it), they will do it. It may also be that as figures deeply embedded in the region's collectivist Labour culture they feel that the region's self image can adapt while preserving its fundamental aspects – and indeed they, as individuals may feel less need to embark on quests for identity than socially and geographically mobile academics and playwrights.

The Campaign for a Northern Assembly presents a more complex case. The CNA has not sought to use the market oriented identity project in its favour. This is no great surprise. The CNA's membership is full of people who have reasons to attack the market-oriented identity project: academic researchers who are aware of the gap between the hype of the new North East and the reality and see it as their duty to highlight this; Labour councillors who mistrust the ideological undercurrent of what their regional leaders agree to with their business associates on the NDC board; ordinary people who feel that their identity is being taken away from them.

However, the CNA has at times shown a willingness to flirt with the community identity project. Its first major publicly oriented document, *Governing Ourselves,* contains a section on the cultural need for regional government which proclaims that

> We are an essentially a working class culture with working class values. The solidarity built on adversity of the mining and shipbuilding communities has its threads running throughout our region .... Because our culture is that of the working class it has always been challenged by the dominant "English Middle Class" culture and occasionally held up to ridicule .... The mainstream [regional] culture, built on the mass workplace, on a strong sense of community, and a high level of social activity is particularly threatened by the very loss of mass workplaces, the imposition of values that put individuals first and even deny the very existence of community and social conditions which drive people to spend more time in the home .... We can start to meet this challenge by learning to value ourselves as northerners. We must realise that the confidence that will be regenerated by rebuilding a strong regional identity is an essential part of our campaign for constitutional, social and economic change in the region.

For many people the regaining of that confidence would be as liberating as the possible economic and social gains of a regional government (Price, 1992, pp.7-8).

However, the CNA has tended to marginalise cultural issues since. Its 1994 public statement, backed by 275 signatures and published in the *Journal*, emphasised economic decline as a result of decisions taken by a government in Westminster that the region's people had voted against as the main reason to demand regional government. This concentration on democratic deficits and central government imposing unpopular and ineffective policies on the region remained the CNA's main theme until the end of the Conservative Government. Its detailed 1996 proposals on regional government (CNA, 1996) made little mention of culture and identity. The CNA as a body has concentrated on making economic and political arguments, not cultural ones, and on making them to the media, and in the hidden world of Labour (and Liberal Democrat) party lobbying.

There has also been relatively little interaction between those most involved in the Regional Community Identity Project and the CNA. Although one contributor to *Geordies* was an office holder in the CNA for a time, the groups have remained distinct. We can put forward several explanations as to why this is so.

One argument is that it is simply not in the CNA's interests to utilise cultural and identity based arguments for regional government for, as Birch (1989, p.67) bluntly explains:

> romantic nationalists are few in number. They can create minority nationalist movements and keep them alive, but they cannot win widespread support in their community unless they can point to broken promises, material disadvantages suffered, or the prospect of tangible gain.

It could be argued romantic regionalists are likely to be even fewer in number, so autonomist regionalists attempting to build support for regional government find it in their self interest not to get dragged into cultural issues. However, interviews and observation of CNA members suggests that they may not see things so bluntly – not least because of the intimate connection between economic and cultural issues in the North East makes it an area where it is possible to describe the closure of the pits as a "cultural crime" (Byrne, 1995, p.71)

My reading of autonomist regionalists in the North East is that they do believe in the power of regional identity – not least because of the way in which it informs their own belief. The artistic, economic and *political* success of the Treasures of Northumbria exhibition in 1996 showed what could be achieved by CNA officers in positions of power and with allies in the region's cultural bureaucracy, but more widespread use of culture as a mobilising tool has been prevented because of two factors: distaste and uncertainty.

The distaste factor is a perfectly reasonable one given the experiences of Bosnia, or even Scotland and Wales. The SNP and Plaid Cymru are models of inclusivity, their nationalism of the civic, not ethnic type. However, even here we see cottage burning in Wales and 'settler watch' in Scotland. Even David Byrne, a left wing romantic regionalist if ever there was one, has been keen to disassociate his plans for regional government from 'nationalism' and its attendant evils. Indeed, given the fact that a 1995 survey of the CNA membership found that about half of its membership were born and grew up outside the region, it is hardly a body which would want to encourage a 'Northumbria for the Northumbrians' attitude.

It is also clearly the case that most active political regionalists, whether in the CNA or not, tend to be socialists or liberals first – and regionalists second. It is highly likely that sections of the political left have become attracted to decentralisation for similar reasons to Scotland. The North-South economic and political divide of the 1980s no doubt convinced many in Scotland that socialism could command electoral support in collectivist minded Scotland but not in individualist, selfish England. Socialism, however, could be obtained in an autonomous Scotland in a social (or even Christian) democratic Europe. As the Scottish left has lost faith in "Britain" so some figures on the North East left have lost faith in England. If the South could be cast off, the decent, unselfish, community minded, Labour voting North East people (along with the Conservative voting but paternalistic and unthatcherite regional Conservatives) could achieve socialism in one region. As Bryne (1992, p.50) put it in *Geordies*:

> The kind of society that people in the North East want and vote for is quite easy to describe. It is more egalitarian, open, has a high level of public services, and is based on full employment in a successful and modern industrial structure. It looks very like Scandinavian social democracy, and the North-East with its 3 million people would make a very reasonably sized Scandinavian nation-state.

The goal is the type of society, and for socialists, with their internationalist leanings, the society should ideally be achievable on a global scale. So for all the shock that hit the people who are now leading regionalists in the 1980s when election results and their own travels revealed that the social democratic collectivism of their families, friends and villages was despised or feared 'down south', they are still socialists first and regionalists second. The old socialist internationalist perspective makes it difficult for such people to lower themselves to appeal to arguments based on regional identity. Culturally oriented figures are aware of this too, which may explain why many of the intellectuals behind the regional community identity project have maintained a discrete distance from the CNA.

The uncertainty factor is related to the paucity of regional 'high' culture. In Scotland, the core symbols of Scottishness are in place, the identity diffused among the people. A Scottish civil society existed which gave the Scottish equivalent of the North Eastern Regional Community Identity Project originators the stages and pages with which to reform what Scottishness meant, and be sure that their re-formulated imagined community would pass to the population through the Scottish press, education system, theatre, music scene, journals and so on. Not only is the Regional Community Identity Project in the North East at least a decade behind its Scottish equivalent – it also lacks the infrastructure of civil society to give those who originated it and want to believe it, or use it politically, any certainty that such ideas have percolated to the people.

Related to this is the sense that the regional identity building of the cultural 'left' in the North East is still under construction. 'Northumbria' has only just superseded 'North East' / 'Geordie', the rural and suburban present and pre-industrial past are only now being slotted into the regional myth-symbol complex. The regional community identity project is still heavily Tyneside centric, and arouses fears of 'Geordie' dominance on Teesside and Sunderland. Added to this, the direction that regional culture is evolving bottom-up in the wake of the collapse of heavy industry is still uncertain. Economic efficiency and democratic accountability are arguments which have their intellectual foundations firmly in place.

The future uses of identity in regional politics

We will end by (perhaps rashly) making a few predictions on how the role of identity in regional politics may change over the coming years. The first key point is that bottom-up changes in regional and sub-regional identities will continue to occur in reaction to social, economic and generational change. What these changes will be is very difficult to predict.

Second, events outside the control of regional actors seem likely to promote the political use of identity projects. New Labour's decision to set up Scottish, Welsh and London assemblies while denying the North East one can only encourage the CNA and other actors to utilise regional identity more fully. The same is true of the protection of privileged levels of public spending in Scotland and Wales. We can argue that Scotland and Wales's advantageous treatment is based on their claims to be distinct nations within the UK. The temptation will surely be for actors in the North East to develop Peter Taylor's thesis that North Easterners are the "Unenglish" (Taylor, 1993) in an attempt to suggest that the North East is distinctive enough to be treated in a special way too. This temptation to maximise the differences between the North East and other parts of England may well extend to the tourism and inward investment marketing teams as they face an ever harder job to attract investment and jobs. Furthermore, Labour's decision to base its Regional Development Agency on the four county North East is likely to accelerate the decline of the "Northern Region" and force those running regional bodies, and the CNA to talk about the "North East": a unit around which a common identity is more likely to emerge and be mobilised.

The economic divide between the North East and the South East which arguably fuelled both the Regional Community Identity Project and the regional government movement looks set to widen again into the next decade. The Government's high pound and high interest rate policy, designed to deal with skills shortages and overheating in the South East, may increase the feeling that the North East really is a place which simply has different interests from the South East. Such feelings can only increase the numbers who see the North East as "us" and the South East as "them".

Finally, we also have to consider the socialising effect that both identity projects have already had. We have to consider how far the ideas of both have percolated into society, and whether the generation currently in school will be leaving it with strong, and perhaps politicised, regional identities.

References

Anderson, Benedict (1983). *Imagined Communities: Reflections on the spread and origin of Nationalism*. London: Verso.

Birch, Anthony H. (1989) *Nationalism and National Integration*. London: Uwin Hyman.

Bryne, David (1989). *Beyond the Inner City*. Milton Keynes: Open University Press.

Bryne, David (1992). *What Sort of Future?* in Colls, R. and Lancaster, B. (ed). (1992). *Geordies: Roots of Regionalism*. Edinburgh: Edinburgh University Press.

Campaign for a Northern Assembly (1996), *Democracy and Progress: Towards Regional Government for the North*. Newcastle upon Tyne, Campaign for a Northern Assembly.

Colls, R. (1995), What is Community and how do we get it? A message to the Member for Sedgefield, *Northern Review* 1, 19-26.

Colls, R. and Lancaster, B. (ed). (1992). *Geordies: Roots of Regionalism, Edinburgh)*: Edinburgh University Press.

Fawcett, C.B. (1919). *Provinces of England: A Study of some Geographical Aspects of Devolution*, London: Williams and Norgate.

Hudson, Mark (1994). *Coming Back Brockens: A Year in a Mining Village*. London: Jonathan Cape.

Hudson, Ray (1991). The North in the 1980s: new times in the "Great North" or just more of the same? *Area* 23(1), 47-56.

Lancaster, Bill (1996), Editorial: Northumbria, *Northern Review* 3, 1-5.

North of England Assembly of Local Authorities (1995). *Regional Government: Consultation Paper*. Newcastle upon Tyne: North of England Assembly of Local Authorities.

Northern Region Strategy Team (1977). Strategic Plan for the Northern Region: Volume 1 - Main Report. London: HMSO.

Price, Don (1992). 'A Northern Culture' in Campaign for a Northern Assembly, *Governing Ourselves*. Newcastle: Trade Union Studies and Information Unit.

Rowntree Research Unit (1972). *Some Inconsistencies in Regional Development Policy: The Case of the Northern Region*. Durham: University of Durham Department of Sociology and Social Policy.

Senior, Dereck (1965). The City Region as Administrative Unit, *Political Quarterly* 36, 82-91.

Townsend, A.R. and Taylor C.C. (1974). *Sense of Place and Local Identity in North East England* (North East Area Study Working Paper 4). Durham: Durham University.

Townsend, A.R. and Taylor C.C. (1975). Regional Culture and Identity in Industrialized Societies: the Case of North East England, *Regional Studies* 9, 379-393.

Wilkinson, Sue (1992). Towards a New City? A Case Study of Image-Improvement Initiatives in Newcastle upon Tyne in Patsy Healey (ed)., *Rebuilding the City: Property Led Urban Regeneration*. London: E and F Spon.

Williams. Raymond (1961). *The Long Revolution*. London: Chatto and Windus.

6 Pride and Prejudice: Two Cultures and the North East's Transition

PETER FOWLER, MIKE ROBINSON AND
PRISCILLA BONIFACE

Introduction

The economic fortunes of the North East of England have long provided a
wealth of material for examination (see for example: Allen *et al.*, 1957;
Smailes, 1968; House, 1969; Warren, 1973; Robinson, 1988; Benyon *et al.*,
1991; Evans, 1995). While this tradition of evaluation is an important one,
however, relatively little attention has been given to the relationships which
exist between economic change and regional culture (but see Williamson,
1992, p.165). This chapter is predicated upon the reality of such a
relationship. We argue, moreover, that it is a two-way relationship
whereby cultural shifts (as something more substantive than a series of
social 'problems') emerge from the process, and that culture has an
influence over the direction, pace and nature of changes within the
economy (Boas, 1928; Rubenstein, 1990; Robinson, 1998).

There has been some focus upon certain elements of North East
culture as they interact with economic change, particularly those which
relate to the region's (working-) class base, its political expression, and the
coal-mining industry (Dennis *et al.*, 1956; House and Knight, 1967;
Bulmer, 1978). Pickard (1989), in his neo-ethnographic study of 'life' in
the declining shipyards of Wearside, provides insight if not analysis of the
culture there, representative of much elsewhere in the region; but, notably
since the demise of coal-mining and thus of the associated and apparently
discrete units of sociological and anthropological study, cultural aspects of
the region have not commanded much serious study. Although some work
on the shift in gender relations resulting from de-industrialization and
changing working practices has been undertaken (e.g. Hudson, 1989), in
the main not enough attention has been given to the cultural dimension of
the region's economy. In particular, there has been a failure seriously and
creatively to integrate socio-anthropological frameworks with economic
analyses. Perhaps this results from a perception of such frameworks as
nothing more than an externality and as a field of study both contentious

and problematic to research, monitor and measure; but culture cannot be omitted from the processes of economic change.

The term 'culture' is used here in its anthropological sense relating to the inherited values, shared beliefs and learned behaviour of society. To borrow from Raymond Williams (1993), it is the culture of the ordinary that is of both interest and importance – the 'everydayness' which underpins social, political and economic practices. We cannot, therefore, deal with culture in purely artefactual and achievement terms; rather we must look to attitudes and behavioural patterns which are themselves both 'culture' and acculturated.

Two initial issues emerge. The first relates to the extent to which the culture of the North East can be considered as wholly distinctive. The fact that similar claims may be made for Liverpool, Manchester, Yorkshire and Lancashire emphasises rather than negates the reality of cultural distinction. While difficult not to descend to cultural relativism, there is a case for considering the North East of England as a distinct cultural region recognizable from the outside, with an identity accentuated by its peripherality in geographical and political terms. Even within the context of the well-articulated and largely economically focused debate on the 'North-South' divide (see, for example, Martin, 1988; Smith, 1989), the North East can be delineated by its culture, much of which, as seen from the outside at least, emanates from Tyneside and the cult of 'Geordieism'. A key point is that distinctiveness is driven internally by a well-developed sense of self-identity which is a key component in the 'pride' which we refer to here.

The second issue relates to the cohesiveness of North East culture. An internal perspective reveals that culture is itself made up of myriad sub-cultures, interacting with legitimate nostalgia on the one hand and pressures to change on the other. It is from an internal perspective that we can identify cultural variations relating to sub-regional demarcations, notably between the three main centres of urban population (Newcastle/Tyneside, Sunderland/Wearside and Middlesborough/ Teesside), around the urban-rural boundaries which themselves exhibit considerable variation, and between the North of the region (Scottish Borderlands) and the South (North Yorkshire), where distinctiveness inevitably gives way to cultural merging. Externally the clues to, and the subtleties of, cultural sub-divisions mean little. Newcastle and Tyneside are identified not only as the centres of the regional economy and accumulated cultural capital, but as dominating external perceptions of the way of life in the region. Certainly, and despite a large rural hinterland, the urban dimension and the legacy of the industrial era are paramount in defining the culture of the North East. Although we can identify some recognition of regional consciousness prior to the industrial revolution (Jewell, 1994), it was the reality of mid to late nineteenth century industrial achievement which has given rise to the present culture and self-identity.

The central thrust of our argument is that in the on-going economic transition of the North East we need to address the cultural dimension. Specifically the region's economic transition is exhibiting and being influenced by a 'struggle' between two cultures. The first is indigenous, supposedly 'traditional' and is internal, historic, distinctive, proud, prejudiced (close to proud), intransigent and hedonistic. The second is external, post-modern, bland, prejudiced (in the exclusionist sense) and also hedonistic in the sense that it seeks to create opportunities for pleasure. By seeking to understand this struggle and the relative balance which exists between these two cultures we can give deeper meaning to the process of economic transition of the region.

Shaking off the past?

Economic change as a process, and the idea of economic transition in the North East, is clearly nothing new. For most of the twentieth century the region's old industrial strength has been weakening as it was either replaced elsewhere in the world or became redundant. Despite centuries of an ostensibly successful rural economy, with Newcastle, for example, emerging as a notable medieval city based on the wool trade, and pastoral Northumberland still retaining a relatively healthy profile, regional culture is defined by the dominance of large scale extractive and manufacturing industry, in particular coal-mining and ship-building. Though these sectors are now poorly represented in contemporary profiles of the regional economy they continue to exert influence in cultural terms.

The economic history of the North East displays three important features which have conditioned regional culture. First, it is marked by dependency. In urban and rural contexts, the region has long been dependent upon a relatively small number of land-owners and employers (usually one and the same), a handful of key industrial sectors/employers, and little industrial diversification. Large industrial plants concentrated employment and culture simultaneously. The dependency of communities on coal-mining, before and during nationalization, is well documented. But, post-nationalization, what has changed? The community of Ellington, Northumberland, still remains precariously dependent upon the last and privately owned colliery in the North East. And the once new 'sunrise' industrial investments from Japan and Korea, no matter how 'hi-tech', continue to create their own dependency. This long-standing dependency feeds expectation and has been recognized as one of the reasons for subdued entrepreneurship and low rates of business start-ups relative to other parts of the UK. Paradoxically, this dependence has also worked horizontally as well as vertically, resulting in regional community and class solidarity exhibited throughout most of this century (Bulmer 1978; Chaplin 1978).

A second feature of economic change which continues to resonate with regional culture is the actual and perceived external intervention in the economy. Decisions on the structure of the economy and its viability in relation to market signals, changing political ideologies and policy directions have been taken, and continue to be taken, outside the region. The closure of Palmer's shipyard in Jarrow 1933, the subsequent hopes raised and dashed of a steel works planned for Palmer's site, through to the closure of Consett Steelworks in 1980, the miner's strike of 1984/5 and the the subsequent programme of colliery closures, and the closure of the last shipyard on the Wear in 1988, are among many examples which have continually and cumulatively accentuated cultural attitudes amongst those both directly and indirectly affected.

Moreover, intervention to deal with and ameliorate the negative effects resulting from external decisions to close and run-down 'traditional' industrial sectors, also comes from outside of the region. Resources and directed policy from central government, Europe and from further afield have, for nigh on seventy years, sought to guide economic change. Special status and industrial policy zoning, whether conferred by local government, national government or the European Union, itself becomes something to protect and the object of struggle. Perversely, Objective 2 and Objective 5b status in the urban and rural upland parts of the region respectively, are indicators of long-standing economic failure, yet, in recognition of a new dependency they are vehemently protected.

The perception that the region has never been in control of its own destiny has given weight to long-running campaigns for regional government. But, in terms of its manifestation in cultural terms, it has fuelled an insecurity which, on the one hand, spans distrust, dismay and anger and, on the other hand, produces stubbornness and resilience. The region has long adopted the discourse of 'struggle' and has been almost happy to portray itself as badly treated, consistently losing out and in receipt of 'hand-outs'. The Jarrow Crusade of 1936 is probably the seminal moment when these feelings touched a national audience, but over sixty years later the same feelings permeate the attitudes and beliefs of both individuals and organizations across the North East. Whatever the economic circumstances of closure of Siemens and Fujitsu in the summer of 1998, old patterns are being re-inforced.

A third feature feeds the others; it is the suddenness of economic change. Such is not necessarily unpredictable but happens at a speed which does not seem to allow for concurrent changes in values. Closure of large physical structures and the loss of large numbers of jobs, sometimes overnight, fits a catastrophic rather than evolutionary model of economic change. In such a way the term 'transition' is misplaced. The economy of the North East has never been a value-neutral background. Whether dealing with change, fighting for closure, or winning investment, it has been wrought with drama and emotion.

These features of economic change all feed into North East culture. But far from seeking to escape from past events and dramas, there appears to be a cultural need to hang on to them. Drama, disaster and distress are remembered and celebrated. Beamish North of England Open Air Museum may be the most high profile expression of this process of remembrance, but this process is widely diffused within the region and expressed to the rest of the world. Diversionary strategies of the 1990 Gateshead Garden Festival, and the Great North Campaign of the early 1990s have long been out-performed by not only an inability to shake off the past, but a resistance to do so. Popular and national representations of the region's culture play upon all of the above features emergent of economic change: the solidarity and struggle of mining communities in Sid Chaplin's, Alan Plater's and Alex Glasgow's musical drama 'Close the Coal House Door'; the melodrama and sentimentality of Catherine Cookson's novels and more recently their television dramatisations; the portrayal of depression of the 1930s in televisions 'When the Boat Comes In'; then again in 1980s with the attempt to escape from a workless region with 'Auf Wiedershen Pet'; the chronological tales of class culture in 'Our Friends in the North'; and the pride and prejudice in the pithy nostalgia of 'Whatever Happened to the Likely Lads'. These examples and others reveal a region which, despite real and fundamental changes in its economy, captures an uneasy admixture of a pride which is recognized as positive and a prejudice recognized as negative.

Aspects of pride

Pride in one's culture relates not only to the satisfaction and self-worth exhibited toward the outward expressions of cultural achievements but also relates to the values and attitudes which lie behind these. The latter are very firmly anchored to the past where pride has emerged from mere survival. Pride in family, community cohesiveness and solidarity in the face of adversity has been one way of dealing with low pay, unemployment and social deprivation. A sense of unification has come from the recognition of external, distant and non-understanding powers, and in achievement in the face of this. Pride, however, should sometimes be punished, and perhaps there is a punitive element in the "dreadful inevitability about [the North East's] economic decline" (Colls and Lancaster, 1992, p.xiii).

There is more to all this than a mere 'chip on the shoulder' attitude. It stems from a huge and near-ubiquitous industrial culture which developed during an era when, for a time, as in Northumbria's 'Golden Age' around AD 700, the North was pre-eminent in what it did. It was not peripheral and not in the least on the margins on world society. All the characteristics of industrial life permeated, and to an extent still permeate, the culture of the North. An important point here is that because industrial achievement is still, just, in living memory, it is especially easy for it to remain as a source of pride. Beamish North of England Open Air Museum caters for and encourages this, albeit selectively. The museum is part of a mosaic of remembering which taps into a cultural longing. Failure, on the part of the 'rest of the world' at least, to recognise this longing and the learning process behind it are also failing to see the relevance and importance of the local and individual. De Lillo (1997, p.11) delineates the phenomenon in his novel *Underworld*:

> Longing on a large scale is what makes history. This is just a kid with a local yearning but he is part of an assembling crowd, anonymous thousands off the buses and trains, people in narrow columns tramping over the swing bridge above the river, and even if they are not a migration or a revolution, some vast shaking of the soul, they bring with them the body heat of a great city and their own small reveries and desperations.

Northern culture contains a pronounced strain of physicality among its traits. Sport, football and athletics in particular, generates pride, inspiration, admiration and, otherwise rather rare, respect. The 'macho' culture relating to the heavy 'Man's Work' which industry and manufacturing characterize, and the attendant pride in it, can be seen as leading to outmoded and obstructive internal manifestations against templates of transitional progress suggested or imposed from outside. A residual suspicion of 'brainwork', for example, could well be reflected in Government figures for 1997 showing all the North East's Local Education Authorities falling short of the national benchmark (*The Journal*, 18th November 1997, p.25).

Uncertainty, which has pervaded both the industrial and post-industrial North East, has produced a remarkably hedonistic culture which stands in bizarre counter-point to some of the region's statistics of social and economic deprivation. The external designation of Newcastle as the world's eighth best 'party city' illustrates this but it is nothing new (*see* Lancaster, 1992, p.59 on the Bigg Market in the 1890s). William Howitt

(1896) was among many who commented on the apparently strange juxtaposition of the hardship and drabness of the mining communities of Tyneside and Durham with their fondness for 'good living, in which they freely indulge whenever their circumstances will allow them', and 'their dress they often affect to be gaudy' and fondness 'of clothes of flaring colours.' The 'happy sing-song hedonists' Ardach (1979, pp.244-58) saw on Tyneside were of the same breed as are Newcastle's notorious young Bigg Market revellers of today.

This outward and self-conscious display of enjoyment and excess and the patterns of high-spending and conspicuous consumption (Piasecki, 1998), may well be a 'coping strategy' in the face of such historic uncertainty and resultant insecurity. It is not meaningless, but is meaningful behaviour illuminating an underlying characteristic of resilience, borne of an acceptance of receiving continual sudden, hard knocks. In this it relates closely to a pride in the local dialect(s). Often displayed as a reaction to 'outsiders' speaking the 'Queen's English', the dialect which, though historic is also remarkably inventive, has been seized upon as a cultural symbol of self-differentiation.   John Peel, the disc jockey, in a BBC programme about National Service, told how he was first billeted with rough young men he thought were Poles, so 'foreign' was their Geordie tongue. On one level, its unintelligibility communicates a wider symbolism that the outside world does not understand the North East.   On another level, the regional accent and the friendliness of its speakers (coupled, of course, with the relative cheapness of their labour) is proving a useful lever for attracting inward investment in the form of telephone call centres.

In one sense pride, as an expression and as an integral part of the internally constructed and learned culture of the North East, is arguably retreating to become the heritage it draws from. In another sense, it retains a power and a deep rootedness and is having to react against the emergence of an equally pervasive and persuasive external culture which exhibits a number of prejudices.

External prejudices

External perceptions of the North East are themselves historic; a long literary and artistic tradition portrays the North as a foreign country (Alderson, 1968). Images from the industrial revolution invoke colourless metaphors relating to grimness, greyness, dullness and dirt. Perceptions and prejudices continue, and the culture of the region remains to many outside 'foreign' or, at the very least, 'strange'. This was/is more than image; justification for the perception is borne out in government statistics which point to an unhealthy, under-educated, poorly-housed, poorly-paid population.

In his famous *English Journey* of 1933, J. B Priestley – from a Yorkshire perspective, so hardly extravagantly remote – described what he saw in travelling from Newcastle to Wallsend: "mean streets ... Slatternly women ... at the doors of wretched little houses ... screeching for their small children, who were playing among the filth of the roadside" (1994, p.310). That reality has gone but the image persists. Over sixty years later, Danziger's (1996) journalistic account of life in the West End of Newcastle brings out aspects of local pride rooted in the past, football and the rituals of 'nightlife'. But at the same time Danziger's account of the social and economic deprivations of Benwell and Scotswood (Benwell CDP, 1978), follows, inadvertently, Priestley and others in a long line of external gazes focused upon a reality witnessed at first-hand. In so doing, arguably it reinforces a national prejudice which has indirect, if not direct, implications for the regional economy. The reality is difficult for the North East to bluster away.

The internal culture of the region is conditioned to react to such prejudices. Whinging about the English capital city, reluctance to move for work as directed so to do by 'outsiders' – such traits may be masking fears of being seen to be irrelevant and inferior. This in its turn feeds the 'we've been robbed' syndrome – the deep suspicion that whether it be inward investment or football, the win by the opposition was unfairly achieved. There may well be some justification for this suspicion. Peter Hetherington, Northern correspondent for *The Guardian*, reported on the recent industrial closures of the region in an article tellingly entitled 'Tell me the old, old story'. He (1998, p.17) quoted Newcastle-based economic consultant, Keith Burge:

> Part of the current difficulty is that with decision-making in London the decision makers ... see only what is happening around them and base policy on that. In picking up messages of house prices being OK and labour shortages, they imagine the country is like that – whereas provincial England is really suffering.

Economic and cultural transition

In the face of dramatic economic change, but particularly over the past decade or so, there are moves, consciously and unconsciously to displace and replace the internal culture of the region with a 'new', externally-generated culture based on the globally-informed service sector and a service-orientated, standardized culture of consumption. An uneasy merging of consumptive cultures is evident.   The internal culture has moved already in quite significant respects towards the 'middle class mainstream'.   It has been forced to.   It is an economic calculation that two of the largest Marks and Spencer's stores in the UK are in the region.   The Co-op may retain a residual sympathy among older shoppers but it has long been 'tarnished' with the industrial ethos from whence it came.   The great regional consumptive expressions of culture – football, sport in general, drinking, eating and enjoyment – continue but now they are melding and merging with ubiquitous global style.

The new, introduced 'Culture 2' (see Table 6.1) is one of sameness, conformity and comfort.   It is selective, seeking to obscure or remove the old culture without examining the values, beliefs and attitudes formed over 250 years.   Remarkably, in view of its supposed 'newness', Culture 2 accepts two features of 'Culture 1', dependence and intervention. Organizations in the North East have bemoaned cuts in European Structural Funds, and are forced to play the competitive, often divisive, bidding games for Lottery monies and Single Regeneration Budgets.   The Northumbria Tourist Board complains that Scotland and Wales has more money for marketing than the North East.   Economic strategy is decidedly short term, ever looking for a 'big hits' such as Siemens and Fujitsu.   These are the economics of desperation, asking questions about just how different this is from the 1930s and the circumstances of the Jarrow March.   Such historical awareness is part of a Culture 1 which is bound to view with some cynicism the attempts by Culture 2 to rekindle a pride, not in the old manufacturing tradition, but in superficial symbols, landscapes and logos.

This highlights the radical break with the North East's industrial past.   Previously, and as part of the internal North Eastern culture still, there was a direct connection between the regional way of life/culture and between its work and ways of employment.   Outside solutions now are not particularly interested in that social dimension, and without that link creating the distinctiveness of regional culture, the North East has no claim for preference over other aspiring and deserving economic contenders. Entrepreneurs tend to be dismissive of the past; indeed they urge people to forget it, to 'break free of its chains'. Yet without its traditional culture, in a sense the North East has only increased its marginality.

The 'new' culture is difficult to argue against.   Its promises, and visions of economic and 'cultural' regeneration have created a powerful ideology against, it must be said, little opposition.   Against a background of decades of economic disappointment and social failure, it has become difficult to argue against the fact that a derelict riverside, for instance, should be transformed into a marina, visitor centre, and shopping complex. Some would dispute the aesthetics, and the economic arguments may be challenged, but in a de-industrializing context they retain considerable currency. However, it is legitimate to enquire of the cultural impacts of such developments and the ideologies behind them.

It has been remarked that "culture, if it is about anything, is intimately connected with meaningfulness" (Stevenson 1997, p.53). As the region continues to deny its industrial culture and as the 'branch-plant economy' takes further knocks, we see a need at least to turn our attention to what the new culture *means* to the region. Kearns and Philo (1993, p.16) focus on meaning and utilize the term of 'other peoples' in their critique of the processes by which places are being 'sold':

> ... .these 'other peoples' have relationships with the city – or, to be more precise, with the particular city places in which they live, work, rest and play and dream; often the places 'left over' after those with power have chosen theirs – that differ (often quite dramatically) from both the relationship lived by the bourgeoisie and the 'respectable' relationship intended for them by the bourgeoisie. And what we further want to claim is that these 'other peoples' hence possess other attachments to the city that differ from the arguably superficial attachments of the bourgeoisie – those to do with property– ownership and fancy possessions, the surface badges of cultural capital ...

Just as industry impinged on the countryside, so the world outside the region now impinges on the North East. Regional culture (Culture 1) survives yet the values of the external culture (Culture 2) permeate it, bringing with it expectations of how the region should behave and perform. So here are not one but two very strong influences demanding cultural change: cultural colonization itself which will probably produce a hybrid anyway and, not least because of the external financial investment underpinning that colonization, an external pressure of expectation to deliver in market terms. The latter is almost 'culture-free', in that it is neither linked to nor concerned with the well-being or otherwise of local culture.

The ubiquity and sameness of the solutions to the region's ills may or may not work in themselves, but they have certainly diminished the area's distinctiveness. The message is clear: do not be different; be like everywhere else; that is the way to economic prosperity. Be that true or not, the fact is that in important ways, the North East is neither being offered the opportunity to utilize many of the strengths inherent from its culture nor being allowed to capitalize upon its distinctive aspects. In the world of Culture 2, idiosyncracies and non-standard behaviour do not fit the externally desired template and the formula needed to gain inward investment. A methodology of transition, driven by a homogenized global culture of giving assistance, is being applied to the individual circumstance which the North East represents.

Conclusion

Our discussion does not indicate any alliance with a romantic cultural re-invention, nor any denial of the realities of what in some respects is rapid and ubiquitous change.   Culture 2 is indeed present in the region and clearly affecting its people.   But surely a more reflective approach is now required, one which seeks to examine consequences in terms over and above neo-classical economics.

Morley and Robins (1995, p.122) ask "Is it possible, in global times, to retain a coherent and integral sense of identity?" They recognize an "imperative to forge a new self-interpretation based upon the responsibilities of cultural translation;" and they suggest that "Neither enterprise nor heritage culture really confronts these responsibilities. Both represent protective strategies of response to global forces, centred around the conservation, rather than re-interpretation of identities." They then argue that "the struggle for wholeness and coherence through continuity" is merely "romantic aspiration". Reporting Hebdige's 1990 pronouncement that "cosmopolitanism is part of 'ordinary' experience", they take the view that "If it is possible, then it is no longer meaningful to hold on to older senses of identity and continuity" (p.124).

Despite what post-industrial theorists may argue (Robinson, 1988), the North East's cultural attachment is to an economy focused on manufacturing through large production units; and yet the region possesses an economy of which two thirds is made up from the service sector and 95 per cent of all companies employ less than 200 people. The problem is that the region and its way of earning its living are out of synchronisation; culture in the North East is framed by a past which the region may desire but which it no longer has. Forgetting some detail, memory recalls that the manufacturing economy not only provided high levels of employment but also provided social and cultural cohesiveness, a sense of identity (independent of class) and a focus for communication and shared beliefs (Edwards and Llurdés i Coit, 1996).   Of course, it is easy now to over-romanticise cultural consensus and social cohesion, and to forget that depression, degradation and depravity were inherent by-products of an essentially exploitative industrial system.   However, the prospect of progress and a measure of industrial spirit appeared to carry important values.

In some senses the Victorian values of entrepreneurship, self-reliance and philanthropy emerged from, and were suited to, a growing manufacturing economy, not an economy in decline (Morris, 1991). Now there appears little in the nature of the service sector to inspire a gestalt-like shift in cultural values. It has merely provided additional cultural capital. On one view, in the consequential situation of cultural stagnation indigenous values have remained fixed to ideas of industrial success which no longer pertain.

What comes first, economy or culture? The region's people, éven the young, exhibit in general a culture which would mesh better with a manufacturing industrial base – work hard, make things, spend money, don't worry. Yet that impulse has been wasted: "The North-East human and material resources have been squandered because it is invited to share an identity which imagines that the real nation lives somewhere else" (Colls and Lancaster, 1992, p.xiii). More research is needed in trying to understand old and new cultures of the region in order to make a better fit with economic developments. Equally, perhaps economic results would be better, more sustainable, if a job was not just a job but related in some way with community and its culture. An alternative, cynical thought is that, whatever the economy, perhaps it will not have any impact on the region's culture provided those who wish to can have a good time. But what is happening now will, in the long term, result in an increasing lack of distinctiveness, an erosion of identity. On the other hand, it is possible that from the melding of Cultures 1 and 2 will come, not a victory for either but the creation of a brand-new Culture 3, different from the traditional but as distinctive in the context of Culture 2 and its clones as Culture 1 was in the past.

There is already some indication that, despite prolonged effort, the imposed outside approach is not bringing about the desired extent of transition. Perhaps the style and nature of the attempt to change the North East are somewhat against its internal instinct and culture. Can it be, therefore, that a more appropriate way towards achieving better success could be found? And ought the way to be to more in tune with the idea of Culture 1 continuing to be significant in the North East? Dare one even ask whether external solutions should continue to be applied in the same way? Has recent experience shown them to be, as well as not very effective, too damaging culturally as well as sociologically?

There is no doubt that the North East needs to change. The question is how change can allow for the inevitable influence of the facets of pride and prejudice which are still extant in both the North East's and outside cultures. A perpetuation of a lack of cultural resolution seems not to be an option. After all, only a limited transition has in reality so far taken place. The existence, culturally, of seeds for a new concerted attempt have been mentioned; efforts towards change need to be the results of a culture of shared, not conflicting, vision. Leonard, writing for the 'think-tank' Demos (1997, p.45), tried to define a future path for Britain as a whole in a way which is extremely pertinent to our region: "To transform a culture it is essential to begin by reaching agreement around the elements of a new ethos."

We believe grounds for such agreement exist in the North East, if what is now unsuitable cultural pride and prejudice among protagonists will allow them to be seen and developed. The new way needs to allow the region, recognizing but counterbalancing global and standard inputs and pressures, to keep and develop the distinctive identity shaped by its particular and overtly fragile existence. There has to be transition in the North East but a new way of attempting it. The North East should be encouraged to let go of its Culture 1, react to Culture 2, and generate a Culture 3. We cannot define what that third Culture should or will be; but we suggest that the challenge is to make it distinctive.

References

Alderson, F. (1968) *View North - A Long Look at Northern England*, David and Charles, Newton Abbot.

Allen, E., Odber, A.J., Bowden, P.J. (1957) *Development Area Policy in the North East of England*, North East Industrial and Development Association, Newcastle.

Ardach (1979) *A Tale of Five Cities: Life in Provincial Europe Today*, Secker and Warburg, London.

Benwell CDP (1978) *The Making of a Ruling Class,* Benwell Community Development Project, Newcastle.

Benyon, H., Hudson, R., Sadler, D. (1991) *A Tale of Two Industries: The Contraction of Coal and Steel in the North East of England*, Open University Press, Milton Keynes.

Boas, F. (1928) *Anthropology and Modern Life*, W.W. Norton and Co., New York.

Bulmer, M. (1978) 'Social Structure and Social Change in the Twentieth-Century'. In: Bulmer, M. Ed. (1978) *Mining and Social Change*, Croom Helm, London, 15-48.

Chaplin, S. (1978) 'Durham Mining Villages'. In: Bulmer, M. Ed. (1978) *Mining and Social Change*, Croom Helm, London, 59-82.

Colls, R., and Lancaster, B. (1992) Preface. In Colls, R. and Lancaster B. (Eds) *Geordies: Roots of Regionalism*, Edinburgh Univeristy Press, Edinburgh.

Danziger, N. (1996) *Danziger's Britain - A Journey to the Edge*, Harper Collins, London.

De Lillo, D. (1997) *Underworld*, Picador, London.

Dennis, N., Slaughter, C. and Henriques, F. (1956) *Coal is Our Life*, Eyre and Spottiswoode, London.

Edwards, A., Llurdés i Coit, J.C. (1996) 'Mines and quarries – industrial heritage tourism', *Annals of Tourism Research* 1, 341-363.

Evans, L., Johnson, P., Thomas, B. Eds. (1995) *The Northern Region Eonomy: Progress and Prospects in the North of England*, Mansell, London.

Hetherington P.(1998) 'Tell me the old, old story', *The Guardian*, 10 September.

House, J.W. (1969) *Industrial Britain: The North East*, David and Charles, Newton Abbott.

House, J.W. and Knight, E.M. (1967) *Pit Closure and the Community*, Geography Department, Newcastle University.

Howitt, W. (1896) *Visits to Remarkable Places*, Longmans, Green and Co., London.

Hudson, R. (1989) *Wrecking a Region: State Policies, Party Politics and Regional Change in North East England*, Pion, London.

Jewell, H (1994) *North South Divide*. Manchester: Manchester University Press.

Kearns, G. and Philo, C. (1993) 'Culture, History, Capital: A Critical Introduction to the Selling of Places.' In: Kearns, G. and Philo, C. (Eds) *Selling Places - The City as Cultural Capital, Past and Present*, Pergamon Press, Oxford, 1-32.

Lancaster, B. (1992) 'Newcastle – Capital of What?'. In Colls, R. and Lancaster B. (Eds) *Geordies: Roots of Regionalism*, Edinburgh Univeristy Press, Edinburgh.

Leonard, M. (1997) *Britain TM*, Demos, London.

Martin, R. (1988) 'The Political Economy of Britain's North-South Divide', *Transactions of the Institute of British Geographers*, 13, 389-418.

Morley, D and Robins, K. (1996) *Spaces of Identity: global media, electronic boundaries and cultural boundaries*, Routledge, London and New York.

Morris, P.(1991) 'Freeing the spirit of enterprise: the genesis and development of the concept of enterprise'. In Keat, R. and Abercrombie, N. (Eds) *Enterprise Culture*, Routledge, London and New York, 21-38.

Piasecki, J. (1998) reported in *The Journal*, 11 September.

Pickard, T. (1989) *We Make Ships*, Secker and Warburg, London.

Priestley, J. B. (1994) *English Journey*, Mandarin (reprint), London.

Robinson, F. Ed. (1988) *Post-Industrial Tyneside - An Economic and Social Survey of Tyneside in the 1980s*, Newcastle upon Tyne City Libraries and Arts, Newcastle.

Robinson, M. (1998) 'Tourism development in de-industrializing centres of the UK: change, culture and conflict'. In Robinson, M. and Boniface, P.(Eds) *Tourism and Cultural Conflicts*, CAB International, Wallingford.

Rubinstein, W.D. (1990) 'Cultural explanations for Britain's economic decline: how true?'. In Collins B., Robins K. (Eds) *British Culture and Economic Decline*, Weidenfeld and Nicholson, London, 59-91.

Smailes, A.E. (1968) *North England*, Nelson, London.

Smith, D.M. (1989) *North and South. Britain's Economic, Social and Political Divide*, Penguin Books, Harmondsworth.

Stevenson, N. (1997) 'Globalization, National Cultures and Cultural Citizenship,' *The Sociological Quarterly*, 38, 1, 41-46.

Warren, K. (1973) *North-East England*, Oxford University Press, Oxford.

Williams, R. (1993) 'Culture is Ordinary'. In Gray, A. and McGuigan, J. (Eds) *Study in Culture: An Introductory Reader*, Edward Arnold, London, 5-14.

Williamson, W. (1992) 'Living the Past Differently: Historical Memory in the North-East'. In Colls, R. and Lancaster B. (Eds) *Geordies: Roots of Regionalism*, Edinburgh Univeristy Press, Edinburgh.

Table 6.1

CULTURE 1	CULTURE 2
Bigg Market	Party City
CWS warehouse	Malmaison Hotel
Newcastle city centre	Graingertown
The Grey Monument	The Angel of the North
Swan Hunter shipyard	The Tall Ships Race
South Shields	Catherine Cookson Country
Meadow Well Estate	Royal Quays
Wearside	Hylton Riverside
Grainger Market	Metro Centre
The Coast	Sea Life Centre
Stannington	Virtual Reality Valley
Fenwick's store	Jackson's Landing
Hancock Museum	International Centre for Life
Corning Glass Works	International Glass Centre
Hartlepool headland	Hartlepool Historic Quayside
The local chippy	Harry Ramsden's
Fat slags	Working out, gyms
Man's Work	Education and service
The quayside	Newcastle Quayside
An allotment	National Garden Festival
Easington, Seaham	Peterlee New Town

7 The Restructuring of Young Geordies' Employment, Household and Consumption Identities

ROBERT HOLLANDS

Introduction

Much scholarly debate has centred on the transition of industrial localities and cities using general concepts like 'post-industrialism' (Williams, 1983; Kumar, 1995) or more specific processes like 'postmodern urbanization', 'global nodal points' and 'city cultures and post-modern lifestyles' (Featherstone, 1994; Soja, 1995; Landry and Bianchini, 1995). The North East of England is no stranger to these debates (Robinson, 1988; Byrne, 1989; Hudson, 1989; Colls and Lancaster, 1992). On a more practical level, these changes are interpreted and played out by development agencies, city councils, community groups, local academics and the indigenous population, as a debate about the possible future of industrial regions ranging from 'gloom and doom' predictions about continued decline, to rosy and uncritical assertions about the vibracy of our post-industrial landscapes. Journalistic treatments about the city of Newcastle Upon Tyne often reflect this ambiguity. For instance, consider the following headlines and bi-lines:

> For a really good time, call Newcastle: city famed for coal and dole is rated one of the best party towns (Montreal Gazette, 14.10.95).

> Children of the giro economy: Peter Hetherington finds poverty rubbing shoulders with wealth in the trouble-torn north-east (the Guardian, 18.9.91).

> Brave new world: Newcastle-upon Tyne has been called the Liverpool of the north. Its old industries dead, its social traditions dying, all hope seems to be abondoned. But there is a new buzz in the nightlife, the popular culture is the envy of the rest of Britain and even the football team has been winning. Does the city have a future at all? (*The Independent on Sunday*, 4.12.94).

One of the difficulties of assessing a region in transition is in separating out some of the ideological claims — what we might want to hear about the city we live in — from the reality of what is actually taking place. The North East has clearly experienced some dramatic changes from its rise to prominance in the industrial age to its supposed 'popular cultural' rennaisance in the 1990s. For instance, during the industrial revolution the North-East was an internationally renowned economic powerhouse. It was a world leader in shipbuilding, heavy engineering and mining, evidenced by the fact that it was exporting 15 million tons of coal annually just prior to World War I. Despite experiencing periods of economic depression, industrial Tyneside enjoyed unemployment rates of around 2 per cent throughout the 1950s. Even as late as 1971, 40 per cent of the workforce was still employed in manufacturing and primary industries, and unemployment was around 5 per cent (Robinson, 1988).

Since this time, jobs in these sectors have more than halved, unemployment reached double figures in 1981 and, with the exception of two years (1989-90), has remained there ever since, and today more than seven out of ten workers are employed in service industries (Robinson, 1994). By the mid-1990s, this post-industrial region was plagued by a curious paradox. While it continued to have one of the highest levels of unemployment on mainland Britain, according to a U.S. based travel consultant Weismann Travel, its regional centre, Newcastle Upon Tyne, ranked eighth in the top ten cities of the world apparently for the quality of its nightlife.

This chapter[1] sets out to transcend some of the rhetoric surrounding this transformation by examining the impact some of these more recent economic, social and and cultural changes have wrought on young peoples' identities and transitions in work, the home and in the consumption sphere in Newcastle Upon Tyne. As such, it seeks to answer a number of important questions. For instance, how has economic change influenced young adults' transitions into, and orientations towards, work, home and the local community? What role does the domestic household play in shaping coping strategies adopted to deal with a less than favourable financial situation, and how does it contribute to the construction of gender identity? How are the consumption patterns and cultures of young people in the leisure sphere predicated on work and domestic life, and in what ways do they appear to be transcending these? What do these changes mean in terms of the remaking of regional identities?

No analysis would be complete without investigating the impact economic change has had on young adults' identities and transitions. Paramount to this restructuring process is the decline in manufacturing and shift towards service-based employment, changing production regimes organised around the concept of flexibility, an increase in women workers, and the global/spatial reorganisation of capital (Harvey, 1989; Bagguley, 1990). A political economy perspective, sometimes referred to as *economic restructuring,* is absolutely crucial in coming to terms with young people's opportunities and attitudes towards work, shifting household relations and the development of inequalities between different regions in the U.K, not to mentions its influence on changing patterns of consumption and the creation of a variety of different lifestyles.

A second process, which ironically has gone hand in hand with restructuring, concerns the increasingly important role cities play in relation to economic change, cultural consumption and the experience of modernity (Giddens, 1990). Modernity and our contemporary experience of the social world can be likened to the phenomenon of being on a merry-go-round-intensely pleasurable at times, but often moving too fast leaving one to wonder where to get off. The city often reflects this mixed feeling of wonder and joy, anonymity and freedom on the one hand, and danger, vice and disorder on the other. Paradoxically, the city is also home to both the expression of regional identities and feelings of security and belonging, as well as increasingly becoming the primary site of the consumption of more global images, not to mention various 'risk' cultures central to many aspects of youthfulness (Beck, 1992).

The crux of my argument here is that there has been a significant shift in the basis of young people's local cultural identity. In this chapter, I draw attention to a weakening of traditional sites of identity formation for young adults, represented by delayed transitions into self-fulfilling work and a lack of opportunity to set up autonomous households, and contrast this with the growing importance of youth cultural activity to construct a sense of self. This reconstruction, centring around the phenomenon of 'going out', involves both a selective reinterpretation of the previously masculine occupational culture, as well as the creative adoption of new lifestyles. Paradoxically, these changes in economic, domestic and cultural life are slowly beginning to question traditional gender relations and roles, while at the same time provide the basis for reaffirming a new sense of regional affiliation.

Much of the following empirical material is drawn from a broader ethnographic study of young adults' cultural experiences of urban city space and nightlife in Newcastle Upon Tyne (Hollands, 1995).[2] In-depth, semi-structured interviews were conducted with fifteen local women and fifteen young men between the ages of 16-31 (average age 25), and participant observation methods were employed to analyze their leisure patterns and use of the city.

Young adults and the meaning of post-industrial work

The economic and industrial history of Tyneside is critical to any understanding of young adults' cultures and transitions. Seventy years ago the backbone of the North East economy was coal, shipbuilding and heavy industry. The local population was predominantly white (as it still is today, with approximately 96 per cent classed in this category) and patriarchal, with women making up only twenty per cent of the workforce (Robinson, 1988, p.12). As Knox (1995, p.93) argues:

> A regional economy dependent on heavy industries and employing mainly male labour went hand in hand with an enthusiastic consensus adoption and approval of the Victorian domestic model — the man as breadwinner, and the women as wife and mother, working in the clearly defined spheres of home and workplace.

This patriarchal ideology, combined with the development of a strong masculine occupational identity (one of the original meanings of the term 'Geordie' is miner) spilled over into the wider local culture influencing the structure of home life, leisure and community. While mining declined substantially throughout the 1950s and 1960s, there were still 12,500 jobs in coal-mining and quarrying and 22,000 in shipbuilding in 1971 (Robinson, 1988). Four out of every ten jobs in the region were in either manufacturing or in the primary sector, only twenty-five years ago. Knox (1995) suggests, these employment patterns and gender traditions were, in one form or another, to shape the region's identity and history to the present day.

The Tyneside economy of the 1990s is almost unrecognisable in light of this earlier history. The so-called 'Thatcherite Revolution' decimated traditional forms of employment, with 38 per cent of the region's total manufacturing jobs going between 1978 and 1984 (Colls, 1995, p.7). Unemployment, at the time the research was conducted, stood at 12 per cent. Less than 20 per cent of local people now work in manufacturing or in the primary industies, while over 75 per cent are employed in the service sector. Many of the new jobs being created in the service sector are part-time and filled by women, who now make up nearly half of the labour force.

The impact of these drastic economic changes on young adults' identities and transitions has clear implications for employment opportunities. Census figures for Newcastle in 1991 show that just over half of 16-29 year olds are in either full or part time work with an employee, nearly 15 per cent are classed as unemployed and just over 16 per cent are students. Comparable Census figures for this age group twenty years earlier (1971) show that 85 per cent were in employment, while only 6 per cent were unemployed (the remainder being deemed economic inactive). School leavers' chances for work have been dramatically altered. Dennison and Edwards (1988, p.111) note that while in 1977 only 8 out of 255 pupils at a local comprehensive school were unable to find work, the figure had climbed to 134 out of 168 only five years later.

Young adults' changed economic prospects will have an important bearing on their attitudes and orientations to work and education, levels of income and spending, and living arrangements and transitions out of the family household. Furthermore, all of these factors impact upon both the patterns of, and opportunities for, consumption and leisure.

While many feel lucky to have a job, our research reveals that the move towards a so called 'post-industrial' service economy has not provided young adults on Tyneside with either a sufficiently strong work identity to replace that offered by the traditional occupational culture, nor enough money to enable them to set up households separate from their own families. Of respondents in our sample, the average net weekly amount of money earned was £86 (or £125 for just those in employment).

Despite the fact that there may have been some increased employment opportunities for women in Newcastle, which can result in higher levels of sociability and self-esteem, most young females are still limited to work in a narrow range of part-time, low paid servicing jobs, characterised by poor promotion prospects, a lack of training and low status.

In our sample, young women who were economically active were currently working in clerical, caring (nursing), shop work and sales occupations. Jobs held previously since leaving school were also located within a narrow band of what might be referred to as traditional women's work. Furthermore, over half of those economically active were in fact working part-time. These factors, combined with a number existing on student grants and others on benefit, meant that the average weekly amount young women earned in our sample was £70 for the overall group, and £105 a week for those in employment.

Young men's occupational choices and possibilities, while not quite so narrow, are also hampered by higher overall unemployment levels not to mention a dramatic decline in traditional forms of manual labour. Clerical and distributive jobs in the civil service, portering and management employment in the hotel and leisure field, sales assistant in the food industry, painting and decorating and bar work were the main occupations currently engaged in. Again, the existence of part-time work, student grants and unemployment benefit meant that the average weekly amount of money young men earned was £102, with waged workers receiving £144.

Over 50 per cent of the males in employment in our sample stated that they would prefer to be doing a different job in five years time. When asked whether they liked their current work, this was one typical response:

> Doug (L31) No. The main reason, boredom. There is no variation in the work. It is a typical desk job. The kind of work, dealing with child benefit, is not really something that holds a great deal of interest for me.

Interestingly enough, while there was some indication that a small section of young men continued to prefer manual work (see also, Hollands, 1994), most of the male sample expressed a strong desire for more challenging professional jobs. Young women were also committed to having a career and improving their position economically, mentioning work like research, management and self-employment, as desirable jobs.

The key question is whether the regional economy can deliver jobs of this nature in sufficient numbers, or provide the necessary training and education for local young adults to move into these careers. Additionally, the continuing effect of gender segregation at work means that both young men and women continue to lose out economically for different reasons. At the moment, it is clear that post-industrial servicing jobs are neither socially rewarding enough for many young Geordies, nor do they provide the financial means for ensuring independent living arrangements and separate households.

Home, sweet, home?: negotiating work and consumption

Household relations and organisation are crucial to understanding how families and individuals cope with and construct their identities in relation to social and economic change. The importance of the household as both an economic buffer and a source of conflict, as well as a central mechanism for constructing gender identities for young adults should not be underestimated. Furthermore, young people's leisure and consumption practices and expenditure patterns are also incomprehensible outside of this domestic economy.

While there exists some general historical and comparative information concerning young people's movement out of their house of origin, a degree of caution is warranted, as both patterns and circumstances have changed. There are also some important gender considerations and variations to consider. Jones (1995, p.22) states that in the mid-19th century, a large proportion of young people left home quite early (in some cases at the age of 10) to live in either their employers household or board and lodgings, while they worked as either servants or apprentices. In a sense, the transition was not a case of complete independence, but rather a move towards an 'intermediate household'. Within this context there were very few young people who could, either morally or financially, set up a household of their own. As the age of marriage in the U.K. fell from the turn of the 20th century until about 1970, and then began to rise, the transitions adopted by young adults changed.   Many young adults, especially those from working class families, typically left home, married and started families within a short space of time (Jones, 1995).

Yet even within these general patterns, there were significant gender differences. Seventy years ago only twenty per cent of women worked outside the home in Tyneside and "provisions for married women's work were virtually non-existent" (Knox, 1995, p.111). Dennis' work on a northern mining village, conducted in the early 1950s, also provides evidence of the persistence of this traditional gender pattern:

> The pure economic fact of man's being the economic breadwinner for his family is reinforced by the custom of family life, the division of responsibility and duties in the household, and the growth of an institutional life and ideology which accentuate the confinement of the mother to the home (Dennis *et al.*, 1969, p.174).

The separation of these spheres, combined with a fall in the age at which people married, signalled a pattern whereby many young women left employment and moved into an autonomous patriarchal household, similar to that of their own parents. Young men, on acquiring a full wage, were also encouraged to 'court', marry and 'settle down', as head of their own household.

The impact of economic change on young adult's job prospects and incomes has implications for opportunities to move out of the family household. Jones (1995, p.25) highlights the emergence of a new 'intermediate' period again, with young adults leaving, largely for independence rather than marriage, and then returning to their house of origin because of a range of social and financial reasons. Marriage rates amongst young adults on Tyneside support this pattern. While fluctuations in marriage rates also reflect changing values in society, not to mention failing to capture an increase in cohabitation, the rate of decline almost perfectly parallels the dwindling job prospects for young adults in the area. According to Census figures from the last 30 years, it can be seen that marriage rates for all ages groups (between 15-29) have fallen by almost half, with the largest percentage drop amongst the 20-24 year olds (from 42.2 per cent being married in 1961 compared to only 12.9 per cent in 1991).

Recent research has suggested a decline in moral respectability for the institution of marriage and fatherhood, particularly amongst young Northern working class men (Dennis and Erdos, 1992). Our research, however, revealed a relatively high regard amongst young men for their future responsibilities regarding partners and children, even though this commitment was most likely to revolve around traditional notions of being the 'breadwinner' and 'protector'. Surprisingly, local men in our sample were more likely than young women to say they would curb their night-time lifestyle because of marriage commitments (46 per cent as opposed to 27 per cent). The vast majority of local men spoke about future relationships primarily in financial terms - for example, not being able to afford them.

Delayed transitions into marriage and autonomous households is the main coping mechanism young adults on Tyneside have in dealing with a less than satisfactory financial position. On the whole, the parental household appeared to be exceedingly generous in accommodating young family members, despite them sometimes being an increased monetary and domestic burden (Allatt and Yeandle, 1992). However, there are important gender differences to consider with respect to economic power and differential expectations concerning domestic duties.

Previous research suggests that young women's subordinate position economically is at least partly dealt with through their adaptation to particular household arrangements (Griffin, 1985; Wallace, 1987; Hollands, 1990). Only one women in our sample, who happened to be in full-time employment, lived on her own. The majority of young females required a shared living space involving partners, parents, lodgers, students and friends in order to cope economically. Just over half lived in a shared household or with a partner, while 40 per cent lived with parents.

This type of household arrangement was made at a cost for young women and the price for this was generally paid in the form of domestic labour. Over 25 per cent of young women in our sample said that they themselves were primarily responsible for the housework and 80 per cent said it was either themselves or it was shared.

Young women who lived with parents produced conflicting results. One unemployed female living with one parent actually did all the housework, partly because her mother was working, but also because she stated that her mother hated housework and she didn't mind doing it. Other research on young people at home has stated that even young women appear to do very little (Hutson and Jenkins, 1989). There were some young women in our sample who stated that they did very little domestic labour and they were primarily drawn from the group who lived at home with their parents.

However, it was young women with children who suffered most with the burden of domestic labour. Local research has shown that leisure for women in this age grouping is nearly halved due to the time consuming nature of child-care (Blackie, 1993). Interviews with women with children were threaded through with references to a lack of time, the difficulty of work, getting baby-sitters and the need for a break from their children. All of these factors point towards the need for adequate nurseries, workplace and shopping centre creches to ensure that young women can take equal advantage of education, training, leisure and job opportunities.

Due to similar economic circumstance many young men also chose to stay at home. While a higher percentage of our male sample were able to afford to live on their own (20 per cent) when compared to young women (only 7 per cent), half of them also stopped with parents, with 20 per cent sharing accommodations (in contrast to over 50 per cent of young women sharing). The virtual collapse of well paid male manual work in the locality has resulted in an increasing proportion of young men staying at home into their late 1920s and even early 1930s.

A number of employed young men living with parents, or a single parent, were able to offset domestic labour by paying 'keep' or board. Overall, only 13 per cent of the men stated that they were primarily responsible for the domestic tasks, while nearly half said it was someone else, usually their mother. Forty per cent, however, stated that the domestic work in their household was shared and this involved both themselves as well as fathers and brothers. While there is still marked gender segmentation within this more flexible domestic labour situation, there are some embryonic examples of changing individual and family forms (Wheelock, 1990).

Economic change has had an effect not only on households, but on local communities in the North East. Many of those based on single industry and occupational homogeneity have declined dramatically. Beatrix Campbell has highlighted the inwardly destructive response of some of these marginalised communities, particularly with reference to the conflict between displaced masculinity and men's involvement in criminality, and women's attempts to reassert values of caring and togetherness (Campbell, 1993). Our research shows that while local communities are certainly not dead, they do appear to be less important to young adults (Hollands, 1995). For example, while neighbourhood friendships remain, few young people, particularly young women, choose to regularly socialise in their local area.

To conclude, while young women have made some advancements economically, they do not appear to be sufficient to offer either household independence or relief from doing much of the domestic work. Young men, due to their slightly stronger financial position, appear to use their economic and social power mostly to avoid having responsibility for housework (see Hollands, 1990), although there are some examples of changing domestic arrangements. Yet household identities remain largely invisible, and gender relations played out here, while important, remain confined to the private sphere. The interesting question is how some of these shifting economic and domestic relations are beginning to impact on young adults' identity formation in the more public sphere of leisure and consumption.

From shipyards to nightclubs: youth cultural identification in the post-industrial city

If work and traditional communities are declining as sources of stable identity, and home life remains a rather private space, where do young adults express what and who they are? A number of theorists have raised the idea that young people are currently reacting to these changes by developing postmodern identities in relation to youth culture and the global media (Redhead, 1993). Yet it is clear that economic change and household re-adjustment have been partly responsible for the shift towards the search for self in the consumption sphere. Similarly, despite assertions to the contrary, contemporary forms of youth culture continue to be influenced by social relations like class, gender, race and locality (Hollands, 1995).

In an atmosphere of declining occupational and community based identities, the city is an ideal place for young people to reconcile tradition and change and literally re-create themselves. The growing importance of the city as a site of social identity is also connected to a set of additional factors. First, it is a public space in the sense that going out is a visible display of identity. Second, urban regeneration in the North East, in terms of leisure, public entertainment and the redevelopment of clubs and pubs, occurred in the same time frame as the region experienced its most rapid economic decline. And finally, this new consumption space did not readily discriminate against the local population on the basis of age, class or gender criteria, and many young adults responded by appropriating sections of the city by claiming them as their own (Shields, 1991).

The existence of these extended transitions and shifts in the sites whereby young people construct their identities has had several significant and contradictory effects. One of the most important ones has been that the meaning and social context of going out has begun to be transformed from a simple 'rite of passage' to adulthood, towards a more permanent 'socialising ritual' or many young adults (Sande, 1994). In other words, involvement in youth cultures and nightlife rituals have simultaneously become a more central and stable aspect of identity formation, as well as created a space for activities, attitudes and behaviours which are neither youthful nor adult, but somewhere in between (a kind of extended or 'post-adolescence').

Paradoxically, involvement in such cultures can both reflect and stand in for the loss of traditional identities, while also allowing young people to step outside of established roles and social expectations. A second effect has been an increased fragmentation of youth identities and cultures in the contemporary period around issues of class, gender, race, sexuality, politics and consumption lifestyles (Featherstone, 1987). Much of this fragmentation derives from young people's increased exposure to global youth culture through the media. The ritualization of nights out has become young adults' attempt to construct a modern equivalent of 'community', or more correctly, sub-communities.

This movement from rites to rituals signals an underlying change in the meaning and function of going out. Nights out for locals historically were much more closely tied to growing up and adulthood. Traditional rites revolved around courtship and marriage, introduction to alcohol, and integration into the community through the local pub. Drinking and going out in the North East was also embedded in a wider cultural apprenticeship based around masculinity and women's subordination in the home. Young men, in particular, were introduced to alcohol in a community setting by fathers or relatives and learnt to appreciate the taste of ale and acquired the capacity to hold their drink (Gofton, 1983). Their induction into the pub paralleled their transition into manual work and manhood generally. Women's subordinate role within this masculine occupational culture meant that they were largely limited to the private and domestic sphere, with the pub being primarily a male preserve. While there is evidence that some women did occupy some of these public spaces, like streets and pubs, their motivations were largely understood within the confines of domestic or sexual discourses (Common, 1951; McConville, 1983).

Clearly, the significance, meaning and wider social context surrounding youth cultures and going out has changed dramatically in the contemporary period (Gofton, 1986). While certain aspects of these historical legacies survive and young people still initially enter into nightlife activities as a form of initiation into adulthood, it is readily apparent that the necessary social and economic conditions to make this wider transition are no longer secure. The decoupling of these processes has meant that going out can be engaged in *for its own sake* and need not be viewed as an element of growing up. The result is that youth cultural activity has increasingly become an important and more permanent site of identity formation in its own right and is utilised by many young adults as a form of expressing their regional identity, rather than a marker signifying adulthood.

Evidence supporting the view that going out has become a more central and significant aspect of personal identity for some young adults, is reflected in our research findings concerning the frequency of nights out and the percentage of income spent on this activity (see Hollands, 1995). Locals went on nights out to the city centre an average of nearly seven time a month and claimed to spend nearly one-third of their total income on this activity. The shift of meaning and social context is exemplified clearly in terms of some of the obvious changes in nightlife culture, like the fact that it is now city and peer group-oriented rather than generationally transmitted and community-based.

At the same time the changed meaning of going out is obscured somewhat by the continued use of traditional symbols like alcohol (Douglas, 1987) and assumptions that this activity is still motivated primarily by courtship rituals, even though sexual activity has clearly supplanted marriage as the main driving force here. While these remain important aspects of the activity, socialising with friends and being part of a recognisable youth culture or style is increasingly becoming the primary 'raison d'etre' of having nights out. For example, young adults in our sample were nearly twice as likely to give 'socialising with friends' as their main reason for going out, when compared to those who said they went out to get drunk, and going out to listen to music and dance was more highly rated than meeting a sexual partner. Furthermore, while the blockage or delay of transitions has meant that some young adults have sought to revive a version of a traditional local identity, by playing out what it means to be a Geordie on city centre streets and pubs, others have chosen to step outside dominant social expectations and roles and utilise this new found space to create different or alternative lifestyles.

One interesting response by young adults designed to deal with economic restructuring, delayed transitions and the shift towards consumption, however, is to attempt to reinvent what it means to be a Geordie. In other words, for many youngsters, regional identity has less to do with work and industrial production and more to do with consumption in the city. And yet while the forms through which some young locals express themselves on a night out are contemporary, they also contain elements which attempt to reproduce aspects of a collective past (Colls and Lancaster, 1992). For example, if young adults can never be Geordies in a true occupational sense, such an identity can be derived from a selective borrowing of historical images and traits, which are then combined with present day experiences and realities in other spheres. Examples of this come from young adults own recognition of attempts by others to rejuvenate elements of the industrial archetype through the image of the Geordie 'hard man':

> Carl (L18) I seem to think it's a bit personified by some of the lads' drinking in the city centre, this stereotypical Geordie bloke, Geordie town, where you have to be hard, you have to drink twenty pints, you have to have a curry, hit a policeman, then go on the pull afterwards, you know.

While the so called 'hard lads' are a minority response, the symbolic aspects of male Geordie culture extend far beyond the stereotype, despite an increased diversity in male identity. In one sense elements of this reconstruction continue to affect the pattern of nights out for many young men, despite the fact that for the majority it is largely a stylistic ritual. Heavy drinking, sexism, and masculine posturing are all elements of this culture.

At the same time, many misconceptions and much of the hype surrounding violence in the city, often confuses playful adoptions of some of these symbols with an undying commitment to masculine values and anti-social behaviour. For example, in our sample, young local men were in fact less likely to experience or be involved in acts of personal violence than local women. Additionally, there was some evidence of changing patterns of male behaviour in this sphere, with sociability and an interests in music and club cultures increasing. There is no longer a singular model of male identity to choose from.

Shifts in the meaning and social context of going out have also changed fundamentally because of the dramatic increase of women in the city (Wilson, 1991). Young local women in Newcastle go out more frequently, would feel worse if restricted, and spend a higher percentage of their income on nights out than their male counterparts. They were also the sub-group most apt to rate the socialising element of going out as their first priority, and they were also most likely to go out in single-sex groups. Despite the persistence of historical and contemporary barriers in employment, the domestic household and leisure (Blackie, 1993), they have made significant leaps forward in terms reforming their own identities in relation to the city, and have played a central role in promoting the socialising aspect of going out. While young women continue to be influenced by domestic and sexual ideologies, female solidarity in dance and drug cultures (Henderson, 1993) and 'ladies only' nights out are two examples of this changing context.

Clearly, some important changes are taking place with respect to young women's attitudes and cultures (for a comparison see Griffin, 1985). While elements of this shift may be partly related to their movement into the labour market and delayed transitions into marriage, it is in the arena of consumption, solidarity and sexuality that young women have fought for and obtained a degree of freedom (Lees, 1993). Many have discovered that it is both possible and viable to maintain female friendship relations as well as engage in courtship, and in certain circumstances they are consciously choosing to have same-sex nights out where they can enjoy the companionship and solidarity of other women.

The assertion of women's power in this sphere, has begun to challenge traditional notions of subordination, domesticity and sexuality (McRobbie, 1993). For example, young women in our sample were much less likely than young men to say that marriage would curtail their night-time activities (27 per cent compared to 46 per cent for young men). Despite carrying a heavier domestic burden, the vast majority of our female informants refused to allow housework to constrain them from going out. Finally, many local women felt more at ease expressing their sexuality and explicitly set out their own terms of reference regarding meeting someone on a night out, as the following quote suggests:

> Meg (L31) I have a reasonable job, a house and mortgage and a car, so I don't really need a bloke around. If I meet someone its on my own terms.

While not all young women were in this economic position, it is clear that many are prepared to put up with, and deal with, male harassment (nearly three-quarters of our local female sample experienced either verbal and/or physical harassment in the city centre), in order to enjoy a good night out with other women.

In fact, the throwing off and loosening of traditional roles generally has resulted in an increased differentiation amongst young adults and created the conditions for a proliferation of more specific youth cultures. Many young people are utilising extended transitions to explore the possibilities of alternative identities around not only gender (Wilkinson, 1994) and masculinity, but also sexual orientation (Whittle, 1994), involvement in social movements and student sub-cultures, not to mention a range of music, drug and dance based cultures (Newcombe, 1991;

Merchant and MacDonald, 1994; Coffield and Gofton, 1994; Henderson, 1993; Thornton, 1995). While elements of these cultures cross over into orientations towards work, education, and politics, one of the main shifts has been a move away from some of the more traditional and localised attitudes and behaviours towards the consumption of more global images and styles.

The phenomenon of going out is not then a frivolous activity only to be enjoyed, but is increasingly becoming a more central element in the production of contemporary youth identities. The changed meaning and social significance of this activity when looked at in the context of economic restructuring and household and community re-organisation, while creating social upheaval, also hints at future possibilities. In the conclusion, we briefly discuss some of the main implications of the changing economic, domestic and cultural life-world of young adults in the region.

Conclusion

The main purpose of this chapter has been to examine the impact restructuring has had on young Geordies' cultures and identities in work, the home and in the consumption sphere. In summary, it is clear that while jobs and the desire for employment continues to be an important and significant element of young people's experiences in the locality, neither traditional nor so-called post-industrial work appear to be able to provide Tynesiders with a viable identity and sense of the future. Similarly, high levels of unemployment, low pay, and poor career prospects mean that many young adults in the region get by and cope through a reliance on and negotiation with the domestic household. Finally, poor economic prospects, delayed transitions into marriage and declining occupational communities, has meant that the search for a modern Geordie identity has largely taken place in what were once marginal spaces in the sphere of consumption and city/urban life. Not surprisingly, considering the lack of alternatives, it is here, rather than at home or in work that identities are being forged and ritualised and where future possibilities might be considered.

At the most general theoretical level, it is important to begin to bring together economic, domestic and cultural perspectives in order to begin to fully comprehend the nature of generation, locality and identity. For instance, the restructuring perspective provides a useful social and historical context for understanding economic change in a particular locality. On its own, however, it can tend towards a kind of economic determinism and ignore the role of the domestic household and the experience of modernity and consumption as important in the formation of identities (Savage and Warde, 1993). Post-modernist theories, on the other hand, often substitute a kind of cultural or global media-based determinism in explaining contemporary youth cultures (Redhead, 1993). In doing so, they tend to miss the significance of economic and household relations and ignore the continuing impact class, gender, race and local cultures play in structuring young adults' identities and experiences. This chapter has argued that it is important to combine studying youth cultures, styles, and experiences of consumption within the context of economic restructuring feminist theory and household responses. This combined approach can help to illuminate and bridge some of the gaps in our knowledge as to the combined effects of capitalism, place, culture, gender and generation.

The chapter has also touched on a number of more specific debates: the important issue of changing gender relations in regions like the North East; and the question of local identity in an age of global change. The fragmentation and transformation of traditional gender identifies is a major issue tied up with changing work, household, and consumption patterns in areas like the North East. While local women appear to be more at the forefront (Wilkinson, 1993), in terms of their acceptance and desire for change, there are also possibilities for the creation of alternative identities for young men in the cultural sphere. The key issue is whether or not economic and political change is keeping pace with these shifts in patterns of consumption. In terms of young women, what is required is literally an economic revolution with regard to providing them with full-time, varied work and decent pay, not to mention a dramatic rethink on the division of labour within the home. For young men, the issue of employment remains a strong one, particularly in the context of the decline of traditionally masculine forms of manual labour. Equally important, however, is the need for a shift in perceptions about their domestic role, not to mention the need for increased diversification in the male leisure sphere (away from just football and drinking to areas like music and cultural production, for example).

In conclusion, it is important to return to the question about the future of regional identities, particularly for young people, in the context of global economic capitalism and the sharpening of what Giddens (1990) has called the 'consequences of modernity'. Clearly, Geordies are increasingly being forced to define themselves, not just internally or locally, but by outside forces such as global economic and cultural change. This fact has contradictory consequences. While it has worked to widen the question of what characteristics constitutes local identity and who might be included in such a defintion — raising questions not only about the changing position of women, but also issues of ethnicity and 'whiteness' (Bonnett, 1993) — the situation can also be easily misread through a kind of post-modern thinking which often ignores important issues like class and the economy. In this sense, Tyneside should not be written off as an industrial dinosaur, with its proud working class history and struggle resigned to the dustbin of history. The past continues to form and shape its identity, and it should be noted that economic issues like poverty, deprivation and social exclusion are very much alive as we approach the millennium. On the other hand, we also need to put notions of Newcastle, as the 'new mecca' of global nightlife, into context. In reality, the locality contains elements of both its industrial past and a new post-industrial future. Young adult's activities and behaviours provide a bridge between these two worlds. If Colls and Lancaster (1992) are right when they suggest that who Geordies are depends on who they imagine themselves to be, then it is crucial that we begin to listen to and understand the rapidly changing life world of the next generation. The future identity of Tyneside may, to an important extent, be traced through what its young people tell us about themselves.

Notes

[1] This chapter is a revised version of an article which appears in the *Berkeley Journal of Sociology*, Volume 41. I would like to thank the Berkeley Journal of Sociology Collective for their suggestions on the original piece and their editorial comments. Thanks also to the three anonomous referees who commented on an earlier draft of the chapter. Finally, a special thanks to Alan Sande, Helen Carr, Rob MacDonald, Mo O'Toole and Jane Wheelock, who all commented on or contributed to this chapter in some way.

[2] I am referring here to my ESRC funded research Youth Cultures and the Use of Urban City Spaces' (reference number ROO 23 4622) (see Hollands, 1995). I would like to acknowledge the financial assistance of the ESRC and the interviews with young women conducted by my research assistant, Ros Taylor. The key to the taped interviews which follow are: RH- Robert Hollands; RT- Ros Taylor;.... pause; (...) material edited out. The names of all respondents have been changed to protect their identities. The name is followed by an L for local and then a number for the person's age (i.e. Sue (L19) - a 19 year old local woman).

References

Allatt, P. and Yeandle, S. (1992) *Youth Unemployment and the Family*, London, Routledge.

Bagguley, P. et al (1990) *Restructuring: Place, Class and Gender*, London, Sage.

Banks, M. et al (1992) *Careers and Identities*, Milton Keynes, Open University Press.

Beck, U. (1992) *Risk Society: Towards a New Modernity*, London, Sage.

Blackie, J. (1993) 'Women's Leisure Experiences and Their Use of Public Leisure Provision in Newcastle Upon Tyne', unpublished MA dissertation, Leeds Metropolitan University.

Bonnett, A. (1993) *Radicalism, Anti-racism and Representation*, London, Routledge.

Byrne, D. (1989) *Beyond the Inner City*, Milton Keynes, Open University Press.

Byrne, D. (1995) 'What Sort of Future?', in Colls, R. and Lancaster, B. (eds), *Geordies: The Roots of Regionalism*, Edinburgh, Edinburgh University Press.

Campbell, B. (1993) *Goliath: Britain's Dangerous Places*, London, Methuen.

Coffield F. and Gofton, L. (1994) *Drugs and Young People*, London, Institute for Public Policy Research.

Colls, R. (1995) 'Born-Again Geordies', in Colls, R. and Lancaster, B. (eds), *Geordies: The Roots of Regionalism*, Edinburgh, Edinburgh University Press.

Colls, R. and Lancaster, B. (eds) (1992) *Geordies: The Roots of Regionalism*, Edinburgh, Edinburgh University Press.

Common, J. (1951) *Kidder's Luck*, London, Turnstile Press.

Dennis, N. et al (1969) *Coal Is Our Life*, 2nd edition, London, Tavistock.

Dennis, N. and Erdos, G. (1992) *Families Without Fatherhood*, London, Institute for Economic Affairs.

Dennison, B. and Edwards, T. (1988) 'Education' in Robinson, F. (ed) *Post-Industrial Tyneside*, Newcastle, Newcastle City Libraries.

Douglas, M. (ed) (1987) *Constructive Drinking*, Cambridge, Cambridge University Press.

Featherstone, M. (1987) 'Lifestyle and Consumer Culture', *Theory, Culture and Society*, 4, 1.

Featherstone, M. (1994) 'City Cultures and Post-Modern Lifestyles', in Amin, A. (ed) *Post-Fordism: A Reader*, Oxford: Basil Blackwell.

Garrahan P. and Stewart, P. (eds) (1994) *Urban Change and Renewal: The Paradox of Place*, Aldershot, Avebury.

Giddens, A. (1990) *The Consequences of Modernity*, Cambridge, Polity.

Gofton, L. (1983) 'Real Ale and Real Men', *New Society*, November 17.

Gofton, L. (1986) 'Drink and the City', *New Society*, December 20/27.

Griffin, C. (1985) *Typical Girls?*, London, Routledge.

Griffin, C. (1993) *Representation of Youth*, Cambridge, Polity Press.

Hall, S. and Jefferson, T. (eds) (1976) *Resistance Through Rituals*, London, Hutchinson.

Harvey, D. (1989) *The Condition of Postmodernity*, Oxford, Basil Blackwell.

Hebdige, D. (1979) *Subculture*, London, Methuen.

Henderson, S. (1993) 'Young Women, Sexuality and Recreational Drug Use', Manchester, Lifeline.

Hollands, R. (1990) *The Long Transition: Class, Culture and Youth Training*, London, Macmillan.

Hollands, R. (1994) 'Back to the Future? Preparing Young Adults for the Post-Industrial Wearside Economy', in Garrahan P. and Stewart, P. (eds) *Urban Change and Renewal: The Paradox of Place*, Aldershot, Avebury.

Hollands, R. (1995) *Friday Night, Saturday Night: Youth Cultural Identification in the Post-Industrial City*, Newcastle Upon Tyne, Newcastle University.

Hudson, R. (1989) *Wrecking a Region*, London, Pion Press.

Hutson, S. and Jenkins, R. (1989) *Taking the Strain: Families, Unemployment and the Transition to Adulthood*, Milton Keynes, Open University Press.

Jones, G. (1995) *Leaving Home*, Buckingham, Open University Press.

Knox, E. (1995) 'Keep Your Feet Still, Geordie Hinnie: Women and Work on Tyneside' in Colls, R. and Lancaster, B. (eds), *Geordies: The Roots of Regionalism*, Edinburgh, Edinburgh University Press.

Kumar, K. (1995) *From Post-Industrial to Post-Modern Society*. Oxford: Blackwells.

Landry, C. and Bianchini, F. (1995) *The Creative City*, London, Demos in association with Comedia.

Lees, S. (1993) *Sugar and Spice: Sexuality and Adolescent Girls*, Harmondsworth, Penguin.

Lovatt, A. (1994) 'Out of Order: Hyper-regulation in the City', a paper presented at the conference 'City Cultures, Lifestyles and Consumption Practices', University of Coimbra, Coimbra, Portugal, July 14-14, 1994.

McConville, B. (1983) *Women Under the Influence*, London, Virago Press.

McKenzie, S (1989) 'Women in the City' in Peet, R. and Thrift, N. (eds), *New Models in Geography*, London, Unwin Hyman.

McRobbie (1993) 'Shut up and Dance: Youth Culture and Changing Modes of Femininity', *Cultural Studies*, 7, 3, October.

Merchant J. and MacDonald, R. (1994) 'Youth and the Rave Culture, Ecstasy and Health', *Youth and Policy*, No 45 (Summer).

Newcombe, R. (1991) *Raving and Dance Drugs*, Liverpool, Rave Research Bureau.

Redhead, S. (ed) (1993) *Rave Off: Politics and Deviance in Contemporary Youth Culture*, Aldershot, Avebury.

Robinson, F. (ed) (1988) *Post-Industrial Tyneside*, Newcastle, Newcastle City Libraries.

Robinson, F. (1994) 'Something Old, Something New? The Great North in the 1990s', in Garrahan P. and Stewart, P. (eds) *Urban Change and Renewal: The Paradox of Place*, Aldershot, Avebury.

Sande, A. (1994) 'The Use of Alcohol in the Ritual Process', unpublished paper, Norlands Research Institute, Bodo, Norway.

Savage, M. and Warde, A. (1993) *Urban Sociology, Capitalism and Modernity*, London, Macmillan.

Shields, R. (1991) *Places on the Margin: Alternative Geographies of Modernity*, London, Routledge.

Soja, E. (1995) 'Postmodern Urbanization: The Six Restructurings of Los Angles,' in Watson, S. and Gibson, K. (eds), *Postmodern Cities and Spaces*, Oxford, Basil Blackwell.

Stubbs, C. and Wheelock, J. *A Woman's Work in the Changing Local Economy*, Avebury, Aldershot, 1990.

Thornton, S. (1995) *Club Cultures*, Cambridge, Polity.

Tyneside Tec (1993) *Training People...Developing Business: Labour Market Report 1992-93*, Newcastle, Tyneside Training and Enterprise Council.

Wallace, C. (1987) *For Richer, For Poorer: Growing Up In and Out of Work*, London, Tavistock.

Wheelock, J. (1990) *Husbands at Home: The Domestic Economy in a Post-industrial Society*, London, Routledge.

Whittle, S. (ed) (1994) *The Margins of the City: Gay Men's Urban Lives*, Aldershot, Arena.

Wilkinson, H. (1994) *No Turning Back: Generations and the Genderquake*, London, Demos.

Williams, R. (1983) *Towards 2000*. Harmondsworth, Penguin.

Wilson, E. (1991) *The Sphinx in the City*, London, Virago Press.

Worpole, K. (1992) *Towns for People: Transforming Urban Life*, Buckinghamshire, Open University Press.

Young C. and Hollands, R. (1994) 'How Far Has the Northern Region Adopted a Harm Reduction Approach to Dealing with Drug Misuse?', *Youth and Policy*, No. 45.

PART III
ENVIRONMENT AND
COUNTRYSIDE IN
TRANSITION

8 A Sociological Perspective on Air Quality Monitoring in Teesside

PETER PHILLIMORE, SUZANNE MOFFATT AND
TANJA PLESS-MULLOLI

Introduction

In 1996 a leaflet was circulated to thousands of households in Teesside by local authorities and chemical industries (funded partly by the Department of the Environment). Entitled 'Air Quality Today', it has two photographs on its cover, comparing Middlesbrough in the 1960s with the present day. The following extract gives a flavour of its message:

> "Smokey old Teesside?
> Fact: Teesside used to be one of the most polluted places in Britain. In the 1960's it suffered from some of the worst air pollution in the country, due partly to domestic coalburning.
> Fact: it used to be - but not any more!..
> The latest analytical techniques available have shown that our air is as good as other towns and cities in the country — and in many cases a lot better. Take airborne particle pollution for example, which is the most significant local pollutant nowadays. National statistics show that in 1994 Middlesbrough had the lowest reported levels...
> Our biggest air quality problem on Teesside is one of perception."

As the tone and content of this leaflet suggest, air pollution is a sensitive issue in Teesside. The key to this sensitivity, and the main reason why pollution is such a politicised issue, is Teesside's long and continuing history of major steel and chemical operations. These industries have been the cornerstones of Teesside's economy throughout the twentieth century, and have given it much of its distinctive character (Briggs, 1968; Beynon *et al.*, 1994; Hudson, 1989). However, they have also contributed to its pollution. While this recent leaflet acknowledges that the conurbation suffered severe pollution problems in earlier decades, it was not as easy to acknowledge that at the time. The equivocal character of official responses to industrial pollution in the 1960s is captured in the following two passages from the 'Teesplan', a major urban planning initiative of the period. The first passage stresses the similarity of Teesside to other cities, and makes any problem seem an accident of topography and climate.

> Levels of atmospheric pollution on Teesside are high, *but no higher than in other comparable urban areas of Britain* ... But the climate and topography of Teesside combine at certain times of the year to make for periods of much more intense atmospheric pollution ... Thus it will be seen that *a fairly average degree of atmospheric pollution* is temporarily exaggerated, at a critical time of the year, by unavoidable climatic fogs and mists (Ministry of Housing and Local Government, 1971, p.324, emphasis added).

Yet on the next page the problem of industrial air pollution is conceded:

> planning policy should be directed at keeping to a minimum residential and other urban development in the areas liable to temperature inversion and, therefore, atmospheric pollution if downwind from the main industrial sources of pollution (p. 325).

An incident in 1981 shows how readily Teesside's association with pollution touched a raw nerve. An advertisement for Crown Paints in the trade press suggested that exposure to the Teesside air offered the most demanding test conditions that Crown could find for one of their products. The implication that the air along the River Tees was heavily polluted brought an angry response from local authorities and industry, with an official complaint to the Advertising Standards Authority and, after an unfavourable response from that quarter, the eventual intervention of local MPs. As a county Pollution Control Group insisted:

> It is not merely a matter of local pride that caused us to pursue this matter — it is difficult enough already to attract new industry to the area, and nationally circulated advertisements which perpetuate the myth of the grimy North East certainly do not assist the local authorities in their efforts (Borough of Cleveland Pollution Control Group, 1982-83).

It is against this background that we explore some ramifications of historical changes in air pollution monitoring in Teesside. The monitoring of air quality has not previously attracted the attention of social scientists. This is perhaps not surprising. Yet the information obtained from monitoring the Teesside air contributes to an often highly charged public debate about the severity of pollution, its distribution, its changing character, and its sources. This debate embraces local authorities (particularly the Environmental Health Departments or their equivalents), national regulatory authorities (formerly Her Majesty's Inspectorate of Pollution, now the Environment Agency), Teesside's steel and chemical industries, various branches of the health service, the local media, and environmental campaigning groups, as well as the wider public. Our own interest in the subject arose out of involvement in epidemiological studies in Teesside which assessed the health effects of industrial pollution (Bhopal *et al.*, 1998; Pless-Mulloli *et al.*, 1998; Phillimore and Morris, 1991). To some degree the wider debate about pollution and its effects in Teesside persists without reference to pollution data.   For instance, public perceptions are likely to reflect direct personal experience of exposure, the evidence of smell and sight, as well as indirect knowledge through media reports of pollution incidents.[1] Yet the meaning and significance of the information from air quality monitoring is a critical facet of official dialogue, shaping the public education or propagandising reflected so vividly in the leaflet cited at the beginning. In this context, the title of the recent book by Lowe *et al.* on farming pollution, *Moralizing the Environment* (1997), would just as aptly describe the complex nature of the dialogue about pollution in Teesside.   Our concern, then, is with the construction of pollution knowledge — the ways in which the story of Teesside's pollution has been told.

An emphasis on the dialogue around pollution issues may, however, give an impression that while scope exists for rival *interpretations* of the evidence, that evidence itself — the monitoring information — is beyond dispute. Yet the manner in which the data are themselves constituted ought also to receive equal social scientific attention, for the distinction between fact and interpretation is far from straightforward. Decisions about where to place monitoring sites, the criteria used in reaching such decisions, and the ways in which monitoring data are aggregated to make the mass of available information manageable, each influence the data generated, and thereby the scope for interpretation. We shall argue that shifts in monitoring methods from the 1950s to the present are not simply about improved data capture. They have also provided an empirical foundation for generating subtly different accounts of Teesside's pollution. To change the type of data obtained about air quality is also to change the nature of the claims that may reasonably be made about the air. In this chapter we shall therefore link together: the changing character of pollution monitoring data in Teesside; the range of inferences such data make possible; and the ways this knowledge of pollution has been represented in Teesside politics.

But we return to the leaflet 'Air Quality Today'. The tone of this leaflet suggests that while Teesside has had to live with a stereotyped image as a polluted place, the facts today paint an altogether rosier picture. One of the 'facts' often cited is the dramatic fall in levels of smoke and sulphur dioxide since the 1960s, commonly conveyed by images of a sharp downward gradient. Figure 8.1 provides an example of this evidence. Such evidence helps to underpin what we have called elsewhere a narrative of reassurance about Teesside's pollution (Phillimore 1998; Phillimore and *Moffatt*, 1999). Yet we would argue that it does not entirely carry conviction. Behind the comfortable reassurances, there are almost certainly some uncomfortable facts about Teesside's pollution. We are obliged to say 'almost certainly' because it has become increasingly difficult to say anything about one aspect of air pollution: its unequal spread. In this age of information, we now know less than we used to about disparities in pollution *within* the Tees conurbation. This might be of relatively minor significance were industrial operations sited well away from residential neighbourhoods. However, although the trend in recent decades has been to increase the separation of housing and heavy industry, several areas of housing, particularly on the south side of the river around Grangetown and South Bank, remain close to major industrial plants. At the 1991 Census approximately 12,000 people were living within one kilometre of such chemical and steel-related sites. The unequal distribution of air pollution within Teesside is potentially therefore of considerable importance, particularly for public health.

The changing character of pollution monitoring data in Teesside, 1950s to 1990s

Pollution monitoring throughout Britain got underway in the 1950s. The Clean Air Act which followed the devastating city smogs of the early 1950s (most notably in London in 1952) also led to the beginnings of a national framework for monitoring air quality. Coal combustion was the primary concern. Responsibility for monitoring was placed in the hands of local authorities, where it has remained. Their statutory obligation has been to provide the information on which progress towards meeting air quality targets could be assessed. However, responsibility for monitoring has always been quite separate from the authority to regulate pollution and enforce compliance with environmental standards. At the risk of some simplification, local authorities have historically held limited regulatory powers in relation to the smaller industrial sites. The main regulatory powers, particularly where major industries were involved, belonged to Her Majesty's Inspectorate of Pollution (HMIP) until its disappearance in 1996, and replacement by the Environment Agency. The sources of pollution (the points of emission), on the one hand, and its spread (ambient pollution), on the other, have therefore fallen within different spheres of official responsibility.

Serious concerns about air pollution along the River Tees date back to the turn of the twentieth century. Such unease may be followed in the annual reports of the Medical Officers of Health for the different local authorities (above all in Eston Urban District). Regular monitoring of air pollutants in the Tees basin did not start, however, until the 1950s, in line with the country as a whole. The earliest routine monitoring in Teesside concentrated on 'insoluble deposits' and ferric oxide, extracted from grit and dust, in recognition of the problems associated with steel production. By the early 1960s a network of over thirty sites were spread across the conurbation (Cleveland County Research and Intelligence Unit [CRIU], 1975). Monitoring of sulphur dioxide and smoke started a few years later, and on a smaller scale (for example, Eston Urban District had two instruments by the end of the 1950s). By the early 1960s, however, measurement of sulphur dioxide and smoke was also becoming well established. Through the 1960s and 1970s, and into the 1980s, air monitoring revolved around these four pollutants: insoluble deposits and ferric oxide (using deposit gauges), and smoke and sulphur dioxide (using volumetric gauges). By 1980 there were well over thirty monitoring sites for smoke and sulphur dioxide in operation across urban Teesside, matching the network for measurement of grit and dust deposits. Such a number was itself an indirect acknowledgement that air pollution posed particular problems in Teesside. Nonetheless, many of the gauges used for monitoring smoke and sulphur dioxide had entered use at a comparatively late date. Around half were introduced in 1974. Even in the mid-1960s, the network of sites for monitoring smoke and sulphur dioxide was comparatively small: for example, in 1965-66 there were eleven sites around the conurbation.

By the mid-1980s, if not earlier, this monitoring system had had its day. The overall pollution picture had changed dramatically over the preceding 25 years or so. The accuracy of deposit and volumetric gauges was increasingly acknowledged to be suspect at the low pollutant levels which were becoming steadily more typical. Moreover, new sources of pollution and different pollutants were supplanting the position of the earlier generation of highly visible pollutants, based on coal combustion. The older gauges were steadily withdrawn during the late 1980s, leaving only a handful of deposit or volumetric gauges in use into the 1990s, before they too were withdrawn.

In its place was introduced an entirely different monitoring system. The new approach to monitoring allowed for continuous analysis, making hour-by-hour observations feasible for the first time. It also made possible fine distinctions at generally low pollution levels, addressing a major weakness of the old gauges. Furthermore, a much wider range of atmospheric pollutants were made the subject of routine monitoring, including hitherto unmonitored gases. By the end of 1993, levels of the following pollutants were being continuously measured in Middlesbrough as part of the routine monitoring programme: *Small Particles* (PM_{10}):

Sulphur Dioxide; Nitric Oxide/Nitrogen Dioxide; Ozone; Carbon Monoxide; up to 25 Hydrocarbons (Volatile Organic Compounds, such as Benzene); and a range of Toxic Organic Micro-Pollutants (Dioxins and Polycyclic Aromatic Hydrocarbons).

This major shift in monitoring practice reflected national changes, as a countrywide monitoring network was consolidated and modernised. The centrepiece of this national monitoring system has been the creation of what is termed the Automatic Urban Network (which includes Middlesbrough), funded by central government. One consequence of these developments has been to centralise monitoring in Middlesbrough in one location. Neither Stockton nor Redcar and Cleveland (Langbaurgh prior to 1996 local government reorganisation) produce the range of data generated for Middlesbrough at its Longlands College site, but in both local authorities monitoring has also been concentrated in one location. There are thus three sites now for continuous monitoring within Teesside. Cost has been one factor in this centralisation, but not the only one. Whatever the sophistication of the equipment at each location, this is a long way from a network of a dozen monitoring sites, let alone the thirty sites of 1980.

The changes in monitoring strategy and methods since the 1950s which we have described are the outcome of several influences. First, there have been significant developments in the technologies available for measuring pollution. Second, the character of urban and industrial pollution has itself undergone major changes since 1950. Finally, scientific understanding of the public health or environmental risks associated with specific pollutants or chemical reactions has evolved over the last four decades. Each of these factors has had a major impact on monitoring policy. To cite one example, the relative importance of volatile organic compounds (VOCs) within the spectrum of urban pollutants, the capacity to measure these compounds accurately, and the scientific perception of their relative importance, have all changed considerably since the 1960s. Clearly the huge improvements in monitoring technology greatly enhance the capacity to describe and measure the composition of the Teesside air more accurately, while reflecting in the process changing judgements about what it is important to describe and measure. Yet this picture of progress and improvement is not the whole story. For one consequence of these developments has been to make certain kinds of inference or description about the Teesside air harder, even as other kinds of inference become easier to make.

The range of inferences air quality monitoring data make possible

To see the history of air quality monitoring as going through two 'stages' is in some respects a simplification of a more gradual evolution. Yet a comparison of 'then' and 'now' highlights the strengths and limitations of monitoring at different periods. While collection of adequate data for inferences about long-term trends has always been the main objective of monitoring, the earlier system of monitoring in Teesside also assisted observation of geographical variations in pollution patterns, because of the diversity of sites employed. Periodic changes to monitoring sites make it very difficult to follow these spatial variations over time between the 1950s and 1980s; but it is possible to draw conclusions about differentials in pollution across the conurbation on the basis of snapshots at different times (the importance of this possibility in an epidemiological context is illustrated in Bhopal *et al.* (1998), and in Pless-Mulloli *et al.* (1998)). That possibility has disappeared, at least for the time being, with the introduction of new monitoring technology. What has been lost in spatial resolution, however, has been gained in temporal resolution (to mention just one gain). Real time monitoring makes it easier than ever before to identify short-term pollution peaks and troughs, and for the first time to do so at the time they occur.

The gains and losses entailed in the change of monitoring regimes may be illustrated by taking two examples. New monitoring technology has proved particularly valuable for an assessment of the pollution load associated with road traffic, for morning and evening rush hours stand out sharply. It has also aided comparisons between cities in the national network, which was one of the main objectives behind the new developments. On the other hand, the potential to capture spatial variations in air quality associated, for example, with routine emissions from industrial sources has been greatly reduced, if not altogether lost. It has become harder to gauge the contribution of industrial pollution through comparison between monitoring sites close to or distant from industry, upwind or downwind, because there are no longer the monitoring sites for making such comparisons.

Today, measurement of the contribution industrial emissions make to air pollution in Teesside can be most easily inferred from comparison of pollutant ratios (where one pollutant is known to be associated particularly with industrial activities, such as the Volatile Organic Compound xylene). Yet such data still provide no indication of how a specific pollutant may be distributed unevenly across the conurbation. A certain amount of additional, non-routine monitoring has taken place, to supplement the evidence from the main monitoring centres — but the very fact that it is not routinely done has tended to reduce the value accorded to observations, with any variations being hedged around with qualifications (TEES, 1995).

How these changes in monitoring practice affect the inferences that it is possible to draw about Teesside's air pollution is best exemplified using monitoring data from different periods. The early phase, when monitoring concentrated on measuring the fall-out from grit and dust, provided repeated evidence of declining pollution with increasing distance from industry. In Eston Urban District, containing steel and associated coking plants, five monitoring sites showed a marked gradient year by year in what was described at the time as deposits of 'undissolved matter' (see Table 8.1 for 1958-60). Table 8.2 shows a similarly marked gradient between areas, using monitoring information covering the six years from 1962 to 1967. Based on all monitoring sites in Teesside, the classification of locations shown here was one devised by the Teesside local authorities. Table 8.3 presents evidence from the early years of smoke monitoring within the Eston area. This table illustrates the typically wide daily fluctuations in air quality occurring over a three month period in 1967. But more importantly it reveals a sharp underlying divergence in smoke levels between areas.

What is consistent in all these data is that the closer to industrial sites the greater the pollution load. In a report by Cleveland County Council reviewing pollution trends in the ten years to 1973, both the overall improvement in the four main pollutants monitored and the local exceptions to this overall picture were noted (Cleveland CRIU 1975, p. 6):

> In the period 1964-73 there has been a general decrease in all four forms of pollution analysed ... the fall in smoke pollution over the ten years is the most marked ... In contrast to the general improvements ... there remain particular sites where the pollution is becoming worse or where high levels of pollution are still being registered. South Bank is the area most severely affected.

In a similar report reviewing the ten years up to 1981, South Bank was one of the areas still singled out as recording highest pollution levels for all four pollutants (Cleveland CRIU, 1982).

Today, the data from monitoring of smoke and sulphur dioxide, extending back to 1961, are available on the Internet. However, not all sites were included in the Internet archive. We do not know what criteria determined inclusion; but we have noticed that one site to have been excluded is that used for Table 8.3. One effect has been to narrow the apparent historical differentials in air pollution within Teesside.

Descriptions of air pollution in Teesside today, centred on one or two locations, inevitably have a different character. No present-day variant of the tables presented here can be reproduced for any pollutant. On the other hand, the new capacity to capture varying pollution levels over short periods of time is illustrated in Figures 8.2-8.4, reflecting different levels of temporal resolution.   The first of these averages fluctuations in small particles (PM_{10}) over a year, taking 1997. The latter two break this down to contrast two days in August of that year, showing hour by hour changes in pollution load: one day (Figure 8.3) when between about 14.00 and 18.00 a distinct pollution episode occurred, along with a further rise in the evening; the other (Figure 8.4) a day of consistently low readings throughout the twenty-four hours.

However, the following quotation regarding benzene, from the Department of Environment/Scottish Office consultation document entitled *UK National Air Quality Strategy* (1996) is helpful in illuminating the strengths and limitations of current monitoring:

> The availability of two years' data from the continuous hydrocarbon monitoring site at Longlands College, Middlesbrough presents an opportunity to examine benzene levels in an urban background location which is potentially influenced by localised, industrial emissions. The hourly data for benzene shows the clear influence of sporadic peaks of benzene superimposed upon a steady baseline in a manner not found at most of the other automatic monitoring sites. These peaks have been associated with industrial sources and the baseline with motor vehicle traffic. Despite there being significant peak hourly mean benzene concentrations of up to 55ppb [parts per billion], annual mean concentrations are barely different from those reported for other urban background sites.

The commentary goes on to acknowledge that exposure of populations living downwind of significant industrial sources of benzene is likely to be "greater than that suggested by data collected at a distant monitoring site which may only register local point source emissions on certain wind directions."

In other words, the gain in the new capacity to interrelate, in this instance, traffic and industrial benzene sources, is confined to the monitoring location itself. Extrapolation to neighbourhoods closer to, or further away from, industrial sites is little more than guess work, with the acknowledgement that there will be an excess in exposure associated with proximity to industry.

It has to be remembered that local authority monitoring data has been gathered primarily to provide the information to fulfil statutory obligations. First as a way to monitor the impact of the Clean Air Act and, in the last two decades, as a means "to assess how far standards and targets are being achieved" (Air Quality Information Archive), the national policy reflected in local authority monitoring has tended to emphasise aggregate data showing declining pollution over long periods (as illustrated in Figure 8.1).

Concern with variations in pollution, either in time or space, has been a secondary matter. However, in recent years an additional purpose has been added: that is, "to provide public information on current and forecast air quality" (Air Quality Information Archive). The nature of the information available is crucial to this public information function, and in this context short-term or geographically localised variations in pollution levels assume much greater importance. Yet as we have suggested, while temporal variations can now be followed with unprecedented detail, the same cannot be said of spatial variations.

Representations of pollution in Teesside politics

These changes in the availability of information inevitably colour official accounts of air pollution in Teesside, for the monitoring system places significant constraints on the conclusions which may be drawn from the available information. It is at this point — where scientific data feed into a consciously constructed narrative, with the purpose of informing public knowledge and public policy — that the interest for the social scientist becomes most apparent. For the technological sophistication of the monitoring introduced in Teesside in the present decade has accompanied and assisted increasing concern with road traffic pollution, reflecting national priorities. It has not accompanied or assisted comparable interest in industrial emissions. On the contrary, that has become more difficult to assess.

In line with these developments, official efforts in Teesside have gone into emphasising two points in particular about the conurbation's air: first, its good quality overall, reflected in the claim that there is no air pollution problem of any significance; second, that the main threat to air quality comes from increasing traffic pollution. The first claim dates back to the 1970s and even earlier (e.g. Gladstone 1976); the second is a product of the last decade or so. The impact of industrial air pollution is downplayed in this official picture. Our argument would be that the monitoring done at the present time encourages oversight of this form of pollution. In the 1950s, 1960s and 1970s, despite official efforts to downplay such pollution (see, for example, Beynon *et al*, 1994; Gladstone, 1976; Sadler, 1990), the evidence from the routine monitoring programme allowed independent judgements to be made. There is virtually no such evidence to call upon now.

One message conveyed as a result is that Teesside is much like other urban centres in Britain. There can be little doubt that this is a convenient message to be able to convey, for industry and local authorities alike. However, there is a paradox here. If Teesside (and more specifically

Middlesbrough) is at the forefront of national developments in air quality monitoring, it is surely not because it is *like* other places but because it is in crucial respects *unlike* them. What makes it different is the concentration of industrial processes along the River Tees which necessitate the atmospheric emission of toxic gases. It may be argued that there is today no steel or chemical production within Middlesbrough itself, for these industries are nowadays located down-river; and that as a result traffic has no rival as a pollution source in Middlesbrough.[2] Yet such an argument also highlights the gaps in monitoring in Redcar and Cleveland and in Stockton, where steel and chemical production is concentrated, and shows the limitations of an over-centralisation of monitoring capacity in Middlesbrough, the part of the Tees basin now most removed from polluting industry.

This is where the phrase 'moralizing the environment', used by Lowe *et al.* (1997), becomes apposite in this context also. Even data on something inanimate like air quality can be 'moralized', in the sense that they figure in arguments about the sources or causes of pollution, which in turn make reference to implicit political or moral values. The leaflet 'Air Quality Today', which we quoted at the beginning of this chapter, perfectly displays how different issues surrounding pollution are 'moralized' for propagandizing or public education purposes.

> So, do we ever get poor air quality?
> Yes. On about a dozen occasions through the year. Sometimes these occur in the summer when we get PHOTOCHEMICAL SMOG drifting into Teesside from as far away as Europe.
> Despite great care being taken with industrial processes, there are occasions when locally-produced short-term emissions of pollutants do occur ... Thankfully these episodes are quite rare. You might also be interested to know two things:
> 1 Peak pollution levels occur on 5th November...
> 2 Nowadays, much of our local air pollution comes from road traffic ...
> Do you get annoyed when people talk down Teesside?.. If we have wrong perceptions of our area, WE CANNOT BE SURPRISED IF OTHERS DO AS WELL. A wrong perception could mean people do not invest in our area ...
> The way forward involves everyone getting together to challenge wrong perceptions.
> [original emphasis]

While poor air quality in the past is associated with domestic coal-burning (see the extract at the start of the chapter), poor air quality in the present is associated with traffic exhaust fumes, imported pollution from Europe, and even Guy Fawkes night. By verbal sleight of hand industrial emissions are (almost) removed from the picture, though these of course loom large in the assertion that Teesside's biggest air quality problem is one of perception, and in the exhortation to challenge 'wrong perceptions'. By the same token, Crown Paints provoked so much official anger with their advertisement not because they drew attention to Teesside's unique traffic problems, nor its special exposure to European pollution, but because of their tacit reference to Teesside' chemical and steel industries.

The attitudes of people living close to the main industries in Teesside are particularly interesting when set alongside official expectations about correcting the 'wrong perceptions' mentioned above. Individuals draw on their own experiences of pollution from the industries they live alongside.

'Thick orange air from industry, especially bad on a windy day.'

'The smell is like rotten eggs and the clouds of black ash litter your dressing table and your window sills.'

An observable result of pollution from the surrounding industry are the fine speckles of dark dust which gather on the window sills of houses etc. When you clean it away, within a few days it's gathered again. Now if this dust can gather so visibly, as residents you worry whether you are breathing it in and how it, long term, will affect you. I know that many people in the area are concerned for the health of their children and see the pollution from heavy industry as causing asthma in their children and poor coughs in their elderly.'

Yet prevailing attitudes often appear to be ambivalent, reflecting a sense of resignation about the prospect of improvement, coupled with an acceptance that pollution may be a necessary price to be endured for the sake of the work associated with it. Such resigned acquiescence is misconstrued if taken to mean that pollution risks are simply accepted, and regarded as merely a minor issue which pales into insignificance alongside concerns over unemployment or crime. Moreover, new local authority policies to offer environmental information, such as Middlesbrough's publication of daily measurements of air quality, do not necessarily change these attitudes. In the neighbourhoods closest to the main industries people can judge for themselves, and know that their experience is not necessarily reflected in readings from monitoring sites a few miles away. Such efforts by local authorities, although intended to reflect greater openness and willingness to share information, paradoxically end up by underlining the gap between local neighbourhood and town hall. The remarks quoted below reflect something of this personal ambiguity about the risks of pollution. The more diffuse risks for health associated with pollution are weighed against the more immediate risks (for the wider community as well as for individual health) associated with joblessness.

'Common sense tells you that a cleaner environment is essential, but for jobs to be lost would also affect people's health. So which is the greater evil? I just honestly don't know.'

'Well, Middlesbrough Borough Council has assumed a statement *(sic)* saying that the air quality in Teesside is as good as anywhere else in the country. But you only have to go forty/fifty miles away from here. The air quality has improved by a fantastic amount. Since I was a child the air quality in Middlesbrough has really, really improved. But I still don't think it's as good as other places I've been to. But whether you'll ever ever get that, I don't think you would. To be practical about it, with the industry that we've got, I can't see that we'd ever, ever, unless we spent a lot money on washing all the stuff and everything, but, I don't know, it might be physically impossible not to have any pollution at all.'

The history of monitoring in Teesside should not be divorced from its place within the wider context of pollution control and environmental debate locally. If one thread runs through this larger story it is of the protection given to the industries which have been Teesside's lifeblood (Beynon *et al.,* 1994; Gladstone, 1976; Hudson, 1990; Sadler, 1990). Therefore, close public or political scrutiny of atmospheric emissions — an activity served by monitoring — has invariably proved a sensitive issue. Air quality monitoring in any city is now part of a national programme, and Teesside is no exception. Yet local circumstances vary, and the national agenda and national priorities may not necessarily reflect these local circumstances. Teesside is a case in point, for the national focus on the importance of traffic pollution has proved convenient for local authorities and industry alike, making possible the tacit, but longstanding down-playing of the significance of industrial pollution.

To argue that it has been convenient to deflect attention from industrial pollution in Teesside is not to suggest a complete absence of scrutiny and surveillance of industry's emissions. The regulatory concern with breaches of air quality standards or emission standards is clearly focused on industries and their activities. The Environment Agency has imposed fines on companies, like ICI, when they have exceeded authorised emission limits; and indeed has recently been ordered by government to take a stricter approach to enforcement. Such enforcement certainly results in adverse publicity for the companies concerned, and refocuses public attention on industrial pollution. Nonetheless, what we might call the pollution accounting indicated by such breaches hardly features in Teesside's own official depiction of its pollution problems, and is not allowed to detract from the 'good news' presented in an outlet like the leaflet 'Air Quality Today'.

Developments in air quality monitoring are not simply a technical matter, but, we would argue, may be seen as a case study in the politics of knowledge. In efforts to measure and describe the ever-changing composition of the Teesside air, a diversity of sources are routinely invoked, and their relative contributions weighed up. Yet the respective roles of industry, traffic, domestic fuel use, 'foreign' pollution, and cigarettes are not only a matter for empirical measurement (Pless-Mulloli *et al.*, 1998); they also find their way into a kind of political morality tale about individual and corporate behaviour. To put it in a different way, built around the seemingly neutral foundations of air quality monitoring in Teesside is a surprisingly complex and politically sensitive debate about accountability.

Acknowledgements

Most of the quotations in the final section come from interviews carried out by Judith Bush as part of a Department of Health funded study entitled 'Public Awareness of Air Quality and Respiratory Health: Assessing the Impact of Health Advice'. This research has been co-directed by Suzanne Moffatt and Christine Dunn (Department of Geography, Durham University). Remaining quotations come from a survey conducted in 1992-93 on self-reported or perceived health in parts of Teesside and Sunderland, carried out for a wider epidemiological study examining the impact of industrial air pollution on health (TEES 1995; Bhopal *et al.*, 1998; Pless-Mulloli *et al.*, 1998). In preparing this chapter we are grateful to Ron Denley-Jones for critical advice.

Notes

[1] The gap between 'lay', unofficial, public perceptions of pollution and its effects, and scientists' expert assessments of that pollution has attracted growing sociological interest in recent years (see Brown 1992; Irwin and Wynne 1996; Moffatt et al 1995; Nash and Kirsch 1988; Paigen 1984; Phillimore and Moffatt 1994).

[2] However, note also the quotation relating to benzene in the previous section for an interpretation of the balance between the sources.

References

Air Quality Information Archive. 'http://www.aeat.co.uk/netcen/airqual/'

Beynon, H., Hudson, R. and Sadler, D. (1994). *A place called Teesside: a locality in a global economy.* Edinburgh: Edinburgh University Press.

Bhopal, R., Moffatt, S., Pless-Mulloli, T., Phillimore, P., Foy, C., Dunn, C., Tate, J. (1998). 'Does living near a constellation of petrochemical, steel and other industries impair health?' *Occupational and Environmental Medicine,*55:812-22.

Briggs, A. (1968). *Victorian Cities.* Harmondsworth: Penguin.

Brown, P. (1992). 'Popular epidemiology and toxic waste contamination: lay and professional ways of knowing.' *Journal of Health and Social Behaviour,*33:267-281.

Cleveland County Research and Intelligence Unit. (Turner S). (1975). Analysis of air pollution readings for Cleveland, 1964-73. Report for the Committee for the coordination of air pollution and noise. (CR 43)

Cleveland County Research and Intelligence Unit. (Reader C). (1982). Analysis of air pollution readings for Cleveland, 1972-81. Report for the Committee for the coordination of air pollution and noise. (CR 386)

Cleveland County Council. (1982-83). Pollution Control Group, Notes.

Department of Environment/ Scottish Office. 1996 The United Kingdom National Air Quality Strategy. Consultation Draft. London: HMSO.

Gladstone, F. (1976). *The politics of planning.* London: Temple Smith.

Hudson, R. (1989). *Wrecking a region.* London: Pion.

Irwin, A. and Wynne, B. (*eds*) (1996). *Misunderstanding science? The public reconstruction of science and technology.* Cambridge: Cambridge University Press.

Lowe, P., Clark, J., Seymour, S. and Ward, N. (1997). *Moralizing the environment: countryside change, farming and pollution.* London: UCL Press.

Medical Officer of Health, Eston Urban District, 1958 to 1960, and 1967. Annual Reports.

Medical Officer of Health, Stockton Borough, 1962 to 1967. Annual Reports.

Ministry of Housing and Local Government (MHLG), Great Britain. 1969-1971. *Teesside Survey and Plan: Final Report.* Vol 2: Analysis. London: HMSO.

Moffatt, S. Phillimore, P. Bhopal, R. S. and Foy, C. (1995) '"If this is what it's doing to our washing, what is it doing to our lungs?" Industrial pollution and public understanding in North-East England.' *Social Science and Medicine,* 41:883-891.

Nash, J. and Kirsch, M. (1988). The discourse of medical science in the construction of consensus between corporation and community. *Medical Anthropology Quarterly* 14:158-171.

Paigen, B. (1982). 'Controversy at Love Canal: the ethical dimension of scientific conflict.' *The Hasting Center Report,*12:29-37.

Phillimore, P. (1998). 'Uncertainty, reassurance and pollution: the politics of epidemiology in Teesside.' *Health and Place,*4(3):203-212.

Phillimore, P. and Moffatt, S. (1994). 'Discounted knowledge: local experience, environmental pollution and health'. In *Researching the people's health*, J. Popay and G. Williams (eds)., 134-153. London: Routledge.

Phillimore, P. and Moffatt, S. (1999). 'Narratives of insecurity in Teesside: environmental politics and health risks'. In: *Insecure Times*. Vail, J. Wheelock, J. and Hill, M. (*eds*) London: Routledge, pp137-153.

Phillimore, P. and Morris, D. (1991). 'Discrepant legacies: premature mortality in two industrial towns'. *Social Science and Medicine* 33:139-152.

Pless-Mulloli, T., Phillimore, P., Moffatt, S., Bhopal, R., Foy, C., Dunn, C. and Tate, J. (1998). 'Lung cancer, proximity to industry and poverty in northeast England'. *Environmental Health Perspectives*, 106 (4):189-196.

Sadler, D. (1990). 'The social foundations of planning and the power of capital.' *Environment and Planning D: Society and Space*, 8:323-338.

Teesside Environmental Epidemiology Study (TEES). (1995). *Health, illness and the environment in Teesside and Sunderland. A Report.* University of Newcastle Upon Tyne.

Table 8.1 Average monthly deposits of 'undissolved matter' (tons/ square mile) at five sites in Eston Urban District, 1958-1960

Monitoring Site	1958	1959	1960
Grangetown Cleveland House	89	60	97
Grangetown Lanny's	18	16	21
South Bank Labour Exchange	29	23	33
South Bank St. Peter's School	13	11	15
Normanby Crossbeck Convent	10	9	11

Source: Medical Officer of Health, Annual Report, Eston Urban District, 1958 to 1960.

Note: 'Short Analysis', Standard Deposit Gauges. Figures rounded to nearest ton.

Table 8.2 Average monthly deposits of 'insoluble matter' (tons/square mile) across Teesside, 1962-1967

Area	1962	1963	1964	1965	1966	1967
Industrial	22	20	18	20	25	21
Semi-industrial	11	11	9	12	12	10
Residential	6	6	6	8	7	7

Source: Medical Officer of Health, Annual Report, Stockton Borough, 1962 to 1967.

Note: Categories 'industrial', 'semi-industrial' and 'residential' are those used in the source. Figures rounded to nearest ton.

Table 8.3 Smoke measurements at three sites in Eston Urban District, 1967. (Microgrammes/ cubic metre)

Monitoring Site	October			November			December		
	Average	High	Low	Average	High	Low	Average	High	Low
Normanby Clinic	71	129	36	185	333	63	153	328	61
Eston Town Hall	41	92	22	159	356	32	101	329	23
South Bank Albert House Clinic	137	243	90	415	732	118	377	659	153

Source: Medical Officer of Health, Annual Report, Eston Urban District, 1967.

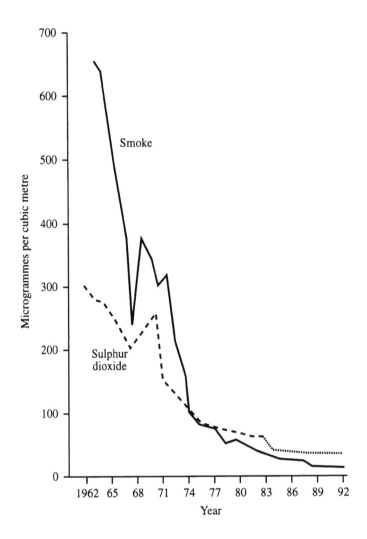

Figure 8.1 Historic trends of SO₂ and smoke in Middlesbrough

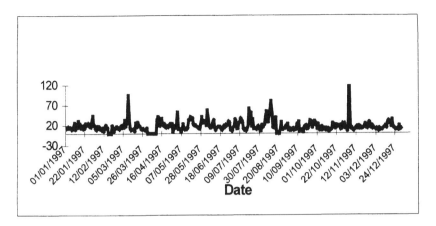

Figure 8.2 24h rolling average PM$_{10}$ in microgram per m^3

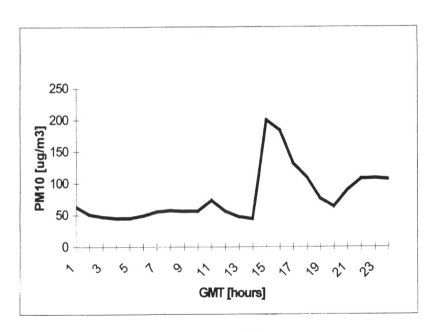

Figure 8.3 Hourly PM$_{10}$ on August 12 1997

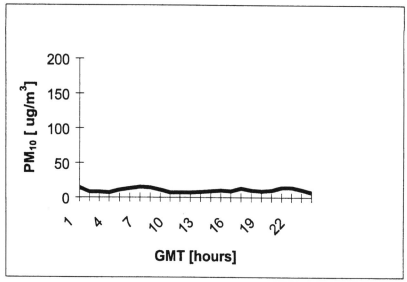

Figure 8.4 Hourly PM$_{10}$ on August 22 1997

9 The 'Rural' in the 'Region': Towards an Inclusive Regional Policy for the North East

NEIL WARD AND PHILIP LOWE

Introduction: The 'Rural' and the 'Regional'

Traditionally, the North East has been seen — both from within, and from outside the region — as predominantly urban and industrial in character. Of course, from the point of view of jobs and population the industrial conurbations of the lower Tyne, Wear and Tees have long dominated the region's economic geography. But interest in, and concern for, their economic prospects has overshadowed the distinctive set of development problems facing the region's rural hinterland.

It is commonplace for rural areas to be seen as 'residual' places in physical and economic planning in the UK — the 'white bits' left over on the map, once the important towns and cities, and their scope for development, have been defined. Similarly, within the academic tradition of urban and regional development studies — in which academic researchers from the North East have for long been eminent players — concerns for the rural dimension to regional policies have tended to be relatively marginal. However, recent years have seen the increasing recognition among local and regional development officials of the particular nature of the rural development challenge in the North East and of the role that rural areas can play in the development of the region *as a whole*.

In December 1997 the Government published its proposals for the establishment of Regional Development Agencies (RDAs) in England which, inter alia, incorporate the rural regeneration work of the former Rural Development Commission. The RDAs began work in April 1999. These institutional changes come at a crucial time for rural policy with reform of the Common Agricultural Policy and European Structural Funds also agreed in early 1999. Drawing on the example of the rural north of England, this chapter critically examines the prospects for an invigorated relationship between rural policy and regional development and argues for further institutional reforms to improve the co-ordination of public policy for rural areas at the regional, national and European levels. What the north needs, it is argued, is a less sectoral (*i.e.* agricultural) approach to public policy in rural areas, and a move to a more integrated and territorially-oriented rural development policy rooted in the links in the region between town and country. In the chapter we consider the nature of rural need and social exclusion, examine the challenges and strengths of the region's rural areas and discuss the new thinking and new institutions required to deliver a truly *inclusive* regional policy which brings benefits across the region's territory.

The new Labour Government has sometimes been accused of being too metropolitan-oriented and unsympathetic to rural concerns. In a House of Commons debate on rural life, Opposition MPs repeatedly made the claim that "the countryside is under siege" and that the Government "doesn't understand" rural issues (Hansard, 4 November 1997, Cols. 176-220). In July 1997 and again in March 1998 over 250,000 people mobilised to demonstrate in protest at a whole host of social, economic and political issues impacting upon rural areas and rural communities in two rallies organised by the pro-hunting Countryside Alliance. Interestingly, Labour now hold more of the 'county constituencies' in Parliament than the other two parties put together, so the charge was more one of perception than the reality of political representation. However, the widespread coverage of the Countryside Alliance's campaign in the media did for a time seem to put the Government 'on the defensive' over rural issues.

We would want to argue that, rather than suffering as a result of the naivety of a metropolitan political establishment, one of the main problems for regions like the North East has been that British rural policy has suffered from a lack of sensitivity to what we call the *differentiated* countryside. As we shall see in the next section, rural areas are on divergent socio-economic development trajectories which are crucially shaped by their regional contexts.

The differentiated countryside and the rural north

Our recent work sits within a wider strand of social science research into rural change that has dealt with 'rural restructuring' (Marsden *et al.*, 1993; Lowe *et al.*, 1993; 1997). This has revealed how different rural localities find themselves on divergent development trajectories as the homogenising pressures of productivist agricultural policy and the post-war consensus around the relationships between state and the market began to break down in the late 1970s. Under such restructuring, localities have been 'freed up' from nationally-derived development strategies such that current development patterns and processes can be seen to reflect new forms of relationship between the 'national, the 'international' and the 'local'.

Counterurbanisation and the urban-rural shift in employment are drawing rural areas into regional patterns of growth and social change. As these trends have become apparent, so we have witnessed the weakening of national agricultural policy and the emergence in agricultural, conservation and rural development circles of an interest in a diversified countryside. Post war agricultural and planning policy both worked with the idea that rural areas were to play a national role primarily for increased food production. Rurality was equated with agriculture and the countryside was the setting for the 'national farm' (see Murdoch and Ward, 1997). Eventually, however, surpluses, budgetary pressures and the recognition that an efficient agriculture may be environmentally damaging brought this agricultural policy into crisis during the 1980s.

With the crisis came a desire to wean farmers off undue reliance on agriculture as their source of income (farm incomes, it was thought, would be continually squeezed). Diversification became a key policy goal, and not just for the farming sector. Rural development interests also encouraged the diversification of rural economies away from their former dependence on primary production. Conservation interests too have sought to reinstate some of the landscape and natural diversity that four decades of agricultural intensification have eroded.

Comparative analysis of the social processes that surround and give rise to the development of land reveals how rural localities are often on diverging development trajectories. If we link the changing function of rural areas (as reflected in these developments in agricultural and planning policies) with the changing nature of regional economies and societies then we can begin to discern an increasingly differentiated countryside. Marsden *et al.* (1993) identified four main sets of parameters as crucial in shaping the development trajectories of rural localities. First are economic parameters, including the structure of the local economy and the role of the state within it. Second are social parameters, including demographic structure and the pattern of social change. Third are political parameters — the organization of politics and types of participation. Finally, fourth are cultural parameters which include dominant attitudes towards property rights and community identity.

The interplay of these different parameters create distinctive patterns of opportunity and constraint, which in turn produce different 'types' of countryside. For heuristic purposes, Marsden *et al.* identified four types: the preserved countryside, the contested countryside, the paternalistic countryside and the clientelist countryside. The typology is not intended to be exhaustive, nor is it claimed that any rural locality can simply be placed exclusively in one of the types. A locality may display a mixture of characteristics (as our discussion of the North East will show). Instead, the typology is intended as a framework which helps draw attention to the differentiation and divergence in the fortunes of rural areas.

The characteristics of the different types were outlined in the following terms. First, the *preserved countryside* is evident throughout much of the lowlands of southern England, as well as in attractive and more accessible upland areas. It is characterised by the dominance of anti-development and preservationist attitudes and decision-making. Such concerns are expressed mainly by middle-class social groups living in the countryside, employed primarily in the service sector, and often working in nearby urban centres. These groups impose their views through the planning system on would-be developers. In addition, demand from these social groups will provide the basis for new development activities associated with leisure, the service sector and residential property. Thus the reconstitution of rurality is strongly shaped by articulate consumption interests who use the local political system to protect their 'positional goods' (Hirsch, 1978; Murdoch and Marsden, 1994).

Second, the *contested countryside* refers to types of formation which lie outside the main commuter zones and may be of no special environmental quality. Here local agricultural, commercial and development interests may be politically dominant and favour development for local interests and needs. They are increasingly opposed by incomers, who may be service class workers or retirees, and who adopt the positions which are so effective in the *preserved countryside*. Thus the development process is marked by increasing conflict between old and new groups, but with no single group yet attaining overall dominance.

Third, the *paternalistic countryside* refers to areas where large private estates and big farms dominate and the development process is decisively shaped by established landowners. Many of the large estates and farms may be faced with falling incomes and are thus looking for new sources of income. They will seek out diversification opportunities and are likely to be able to implement these relatively unproblematically. They tend to take a long term management view of their property and adopt a traditional paternalistic role. These areas are likely to be subject to less development pressure than either of the above two types.

Fourth, there is the *clientelist countryside* which is found in remote rural areas where agriculture and its associated political institutions still hold sway but where farming can only be sustained by state subsidy (such as through less favoured area headage payments, EU regional funds and welfare transfers). Processes of rural development are dominated by farming, land-owning, local capital and state agencies, usually working in close (corporatist) relationships. Farmers will be dependent on systems of direct agricultural support and any external investment is likely to be dependent upon state aid. Local politics will be dominated by employment concerns and the welfare of the 'community'.

In the North East we can observe the interplay between what can be cast under this categorisiation as the paternalistic countryside and the clientelistic countryside. The region still contains a preponderance of large landed estates, including that of England's largest landowner, the Duke of Northumberland, which exceeds 100,000 acres. As well as the large private estates, the region also contains institutional landowners — the Ministry of Defence, the Forestry Commission and the former regional water authority, the now privatised National Coal Board (NCB) and the National Trust.

Throughout much of the rest of England traditional rural landownership has been mainly eroded 'from below' largely through the tenantry becoming owner-occupiers. In contrast, the extensive institutional ownership in the North East represents an external imposition of large-scale public or quasi-public ownership. It occurred largely in the middle period of the twentieth century when private landownership generally was in retreat. In a series of events, large tracts of land in Northumberland were transferred from private estates into public control. These events often co-incided with an estate facing specific financial pressures.

The first example is in 1911 when 19,000 acres (7,600 ha) of the Redesdale Estate was bought by the War Office as an artillery range for the newly formed Territorial Army. Since then, the military's land-holdings in the county have grown, and currently amount to 57,130 acres (23,120 hectares) of freehold and leasehold land (see Woodward, 1998a,b). The bulk of this estate is made up of the Otterburn Training Area, currently used for training troops in the use of heavy artillery, and which comprises almost a quarter of the territory of the Northumberland National Park.

Similarly, the Forestry Commission has acquired large tracts of land in Northumberland. For example, the Commission was able to acquire 47,000 acres (19,000 hectares) of land in the North Tyne catchment in 1932 from the Duke of Northumberland's estate. The sale was a forced one because the estate had to raise death duties following the deaths of two Dukes in quick succession. The transfer included the Duke's former hunting lodge, Kielder Castle, and the land acquired became the core of the Commission's new Kielder Forest. Land continued to be acquired up until 1969, with large parcels obtained from the Church Commissioners, the Swinburne Estates and the Redesdale Estate. By the early 1970s, the forest was fully established, with over 125,000 acres (50,000 hectares) of forest plantations (mainly in Northumberland, but also in Cumbria).

As a result of the transfers of such large tracts of land, we can see what amounts to the 'carving up' of the region's rural landscape between these state and quasi-state agencies and the local landed estates. The geography of landownership by state agencies is very much determined by where the opportunities for land acquisition have arisen. The presence of the National Trust in the county is similarly influenced by the fortunes of landed estates. The National Trust did not enter into those parts of the North East countryside where the big landowners were seen to be secure, particularly in the far north of Northumberland.

For some of these institutional landowning interests — those of the NCB and Forestry Commission in particular — their approach to land management and rural development was, in part, driven by socialistic planning. The Forestry Commission developed whole new villages for forestry workers, such as the isolated village of Kielder. In the 1950s, the village was originally envisaged to house 2,000 families including unemployed people relocated from the industrial conurbations. However, the introduction of the chain-saw meant that even with the expansion of the forest, employment never reached these levels, and the village stands as a monument to an almost Stalinist form of rural industrial development, suffering acute problems associated with peripherality, depopulation and loss of services.

The rural development challenge in the North East Region

Large swathes of the North East region are currently designated as Rural Development Areas. This qualifies them for public funds to support economic and social regeneration. The designation covers 75 per cent of Northumberland and 80 per cent of County Durham and is based on a set of criteria which include persistent high unemployment, low economic activity rates, an over-dependence on a narrow range of economic activities, decline of local services and geographical remoteness (Rural Development Commission, 1993). In such areas, economic disadvantage and social exclusion can often go unrecognised. This is partly a problem of public perception. Rural poverty often exists alongside affluence, or is found in isolated places. It occurs on a much smaller scale and concentration than urban poverty. Indeed, predominant cultural images of the countryside — of quaint villages and sweeping landscapes — leave social and economic problems out of the picture. A second problem in drawing attention to rural need is the fact that many of the indicators conventionally employed by government to measure disadvantage are based on an 'urban' view (Dunn *et al.*, 1998). While a household with a car in an urban area may be an indicator of advantage, a car in a rural area may be a necessity in order to reach work or basic services, and could just as possibly be an indicator of disadvantage in terms of geographical remoteness. In addition, rural poverty can go unrecognised because of an unwillingness of people in rural areas to recognise or accept its existence (Woodward, 1996).

Across Europe, discussion of poverty is being replaced with discussion of exclusion. This shift is more than a mere swapping of words. Where talk of poverty tends to focus on the insufficiency of material resources (cash and jobs), concern for social exclusion focuses attention on the mechanisms which enable participation in social and economic life. In a report to the European Commission, Patrick Commins has argued that there are four complementary processes which are crucial in influencing social inclusion and exclusion: civic integration through the democratic system; economic integration through the labour market; social integration through the welfare state system; and inter-personal integration through networks of family and friends (Commins, 1993). This perspective is useful in understanding rural problems. While poverty often tended to be regarded as a financial inability to participate in economic life, the concept of exclusion is more dynamic, and points to inadequacies with the functioning of a system, instead of seeing individuals or groups of people as mere 'victims'.

Rural development problems are often compounded by two distinctive features of rural areas — their remoteness and their sparsity of population. Relatively low incomes, poor transport links and declining rural services all tend to suffer as a result of these two factors. Some rural areas are far away and difficult to get to, and once there we find too few people to sustain the services and facilities expected by people in urban areas. Northumberland, for example, contains some of the most remote areas in England and has just 61 people per square kilometre, compared to a national average of 376. In the Districts of Tynedale, Alnwick and Berwick, there are fewer than 30 people per square kilometre — less than one tenth of the national averge (Office of National Statistics, 1998).

Recognition of the vulnerability of the North East's rural economy and society culminated in the designation of much of the rural north as an Objective 5b area under the European Union's regional policy. From 1994 to 1999 the Northern Uplands have qualified for some £90 million of EU monies to promote structural adjustment in the rural economy to help reduce the rural region's dependence on agricultural production, improve skills and training services and enhance environmental conservation. When coupled with matching funds from local authorities, national government and the private sector, the programme totals almost £220 million of resources for rural development in the region. This represents an unprecedented package of funding. By way of comparison, the Rural Development Commission only distributed about £30 million in grants anually for the whole of England. However, the European funds are for a fixed term only, and the demands of expansion of the European Union to the east, to take in countries like Poland and Hungary, mean that such large scale programmes for rural areas like the Northern Uplands will not continue at this scale beyond the year 2000. Added to this is the challenge of the next round of reform of the Common Agricultural Policy, agreed in the spring of 1999, will see further reforms to the subsidies paid to farmers from 2000 onwards. European reforms will not see an end to the important funding role of European institutions. They will, however, require new thinking and pose new challenges for how rural development is pursued in regions like the North East. Before thinking about the strengths of the North East's rural areas and what they have to offer to the regional economy as a whole, it is first worth reflecting on the nature of the challenges ahead.

The economic health of the farming sector remains one of the main challenges facing the rural North East. Although farming represents a shrinking contribution to the region's rural economy, it is still a significant employer in some areas, providing as much as 10-15 per cent of local jobs (Ward and Lowe, 1999). But farming in the region remains highly dependent on production subsidies from Europe, and these are likely to drop and eventually be eliminated altogether as European markets are opened up to world trade. A farming community for long reliant on such subsidies will, at least in the medium term, be compelled by economic necessity to adjust to new rules of the game. Fortunately for many farmers in the North East, the general direction of European (and British) agricultural policy is to gradually expand the resources given over to environmental schemes under which farmers receive payments in return for farming in particularly environmentally-friendly ways that enhance the quality of the rural environment, managing valued landscapes and wildlife habitats. The region is well endowed with attractive countryside landscapes, with a National Park in Northumberland, two Areas of Outstanding Natural Beauty and numerous Sites of Special Scientific Interest. Parts of the Pennine Dales have already been designated as an Environmentally Sensitive Area, with farmers there paid to maintain traditional farming systems.

Economic pressures are often claimed by farming groups to be most acute in the hills, with margins particularly low for livestock farms. While the diagnosis would suggest a need for greater efficiencies and competitiveness, there is little evidence that livestock production in the North East enjoys any substantial comparative advantage over its competitors in other regions. Moreover, the marketing of agricultural products remains highly traditional. Only a small fraction of finished livestock bears any form of regional identity which might command a premium on the price, and regional meat procesessors are highly concentrated, with a third of the region's abattoirs handling 95 per cent of the throughput. Over-capacity and the pressures to conform to new standards in the industry are squeezing the meat processing sector. But in any case, meat processing is highly integrated into national markets and does not operate as a regionally-distinct and seperate sector (with the ability to market say 'Northumberland lamb', which might command a better price).

Of course, the most traumatic development for livestock production and meat processing in the region of recent years has been the BSE crisis. National estimates put the losses incurred by the beef industry at £1.2 billion in the first year of the crisis alone, with the north of England pointed to as a particularly vulnerable region (DTZ Pieda, 1998, p. 53).

Added to these structural and food safety challenges that plague agriculture in the region is the looming prospect of CAP reform. The 1999 reforms build on the last round of reforms agreed in 1992 and include further cuts in price support subsidies, coupled with increased direct compensatory payments to farmers. Increased provision is made for environmental conditions to be placed on these compensatory payments and there is to be an expansion in schemes offering environmental payments to farmers. With much talk of transforming the CAP from a Common Agricultural Policy to an Integrated Rural Policy for Europe, an important new feature will be a new Rural Development Regulation to promote rural development and environmental management. The approach this measure embodies could well prove to be the shape of [rural] things to come from Europe.

Most of the agricultural sectors in the North East will be affected by CAP reform, if not through the basic commodity regimes (the sheepmeat regime, for example, is not specifically being altered) then through changes to the rules for payments in Less Favoured Areas. Although it is notoriously difficult to make reliable predictions about the impacts of policy reforms on the agricultural economy, the last round of reforms in 1992 do provide some pointers.

Despite widespread fears and gloomy predictions from farming groups, the last round of CAP reform did not have a major impact on the number of jobs in farming. In fact, the national rate of decline in agricultural employment in the five years after 1992 (a 4.5 per cent loss) actually slowed down, compared to the five years before when the loss was 7.9 per cent. A study by agricultural economists at Reading University argued that employment changes induced by CAP reform have to be seen in the context of wider and longer-term structural and technical changes in farming, such as the steady substitution of full-time workers by various forms of more 'flexible' labour including agricultural contractors and part-time workers (Errington *et al.*, 1996). In any case, because the reform compensated farmers for price cuts, the indirect effects, off the farm, have been greater than the direct effects on the farm. The Reading study estimated that, from the last round of CAP reform, the employment effects in the agricultural supply industries and in abattoirs and food processing amounted to over four times the losses suffered by farming itself.

It is important, therefore, to recognise that CAP reforms will be but one ingredient among many in the economic restructuring of the rural economy, and the impacts are again likely to be greater outside the agricultural sector than within. With farming contributing such a small and diminishing proportion of jobs and economic activity in the countryside, it is important to move away from the idea that the future health of the rural economy is crucially dependent upon the health of the agricultural economy. The two things are not the same, and the latter is a smaller and smaller component of the former. This conclusion only serves to highlight the importance of the new approach to rural development embodied in the new Rural Development Regulation.

The new Regulation will establish an integrated legal framework for farm and rural development and agri-environment measures which have, until now, been provided under several different schemes. Funds are to be allocated on the basis of 7-year integrated programmes drawn up 'at the most appropriate geographical level' within Member States in a similar way to the current Objective 5b programmes under the Structural Funds. This move represents a shift in emphasis and funding responsibilities to make rural policy a more central feature of the CAP. In the longer term it opens up access to a much larger and more flexible budget for rural development measures such as village improvement schemes and community development initiatives, as well as extending the discretion of Member States in implementing the CAP. It also opens up the possibility for a 7-year integrated rural development strategy for the North East region. The reform could signal an end to the damaging separation between agricultural policy and rural development policy in the UK, which have tended to pull in opposite directions. It holds out the promise of fundamentally redirecting resources away from commodity support towards rural development and environmental management in the future. The challenge in the short term is to establish the institutional structures and procedures, both at the national and the regional level, that will facilitate that redirection.

Of course, agriculture is not the only primary industry in the rural North East, and many of the development problems faced beyond the city limits in the region are linked to the long-term decline in coal mining. In Country Durham, for example, the last colliery was closed in 1993 bringing to an end an industry which in 1951 had employed 74,000 people in the county. At the time of writing, only one colliery remains in

Northumberland, and the open cast mines that now predominate in the sector have not brought with them the anticipated jobs. Mine closure poses particularly acute and spatially-concentrated economic and social problems, not least because of the degree of dependency of local communities on the single enterprise.

Turning to the strengths of the region's rural areas, the food and tourism industries are currently the major growth sectors in the rural economy, and both continue to hold significant potential for further growth. The food sector contributes in excess of £1.5 billion to the regional economy and employs over 25,000 people. But over and above its size and significance, the sector also enjoys a number of intrinsic strengths in the North East. The first comes from the region's agricultural complexion. Agriculture in the North East is diverse in its nature, and produces a wide range of meat and arable products. A number of raw materials such as organic foods, livestock products from low-intensity farming systems, sea food, game and certain traditional foods all have a healthy and environmentally-friendly image and the potential, given modern consumer trends, for 'capturing' more of the value-added for the region's firms. A second strength is the structure of the sector, with a good critical mass of small and medium sized enterprises as well as larger subsidiaries of national and multi-national firms. This structure means that training and infrastructure needs for the sector are well-met. Moreover, food sector SMEs are responding well to new opportunities for growth in supplying higher value-added, specialised food products orientated to either local or niche markets or to supplying the large multiples.

Another potential strength for the rural North East lies with tourism. Leisure and tourism have long been pointed to as an opportunity for hard-pressed farmers scraping a living from crops and livestock. In many regions there has been a risk that their potential is over-played. How many golf-courses and bed and breakfasts can a rural area sustain, farming groups have been quick to ask. However, the rural North East is one region where the potential for future growth continues to be quite marked. The annual value to the regional economy brought by tourism amounts to over £800 million (North East Regional Development Agency Contact Group, 1998, p.47). For rural areas the benefits can be wide-ranging. Not only does a steady stream of visitors bring jobs and wealth to a rural locality, but also the spending power of tourists can mean the difference between economic viability and closure of essential local services. Tourism can thus help support and justify vital infrastructure. Tourism promotion also makes a crucial contribution to the marketing of the region as well as to the development of particular sectors such as the food sector. By helping to upgrade leisure facilities and the environment, tourism can more generally contribute to the quality of life in the region. Appropriately developed, it can promote the conservation and enjoyment of the region's natural and built heritage.

The Regional Development Agencies and rural development

For the last 15 years, it has been commonplace to suggest that the countryside is 'at a crossroads', with an old order crumbling, to be replaced with a new one. However, never has this claim been more pertinent than the current period. The election of the new Labour Government served to provide a fresh political opportunity for a strategic rethink of rural policy and came at a time of impending European reforms. Crucial amongst the Labour legislative programme has been the plans for regional devolution and the establishment of the Regional Development Agencies (RDAs), drawing staff and resources from English Partnerships, the Rural Development Commission and the Government Offices.

The RDAs have responsibility for rural development as it has been conventionally understood in England. They are also to take "a leading role" on EU Structural Funds (Department of Environment, Transport and the Regions, 1997, p.44). The Government has reaffirmed this point, in responding to the Select Committee on Environment, Transport and the Regions, by emphasising that "the Government is committed to promoting the interests of rural areas and believe that this can best be done by addressing their particular needs within an overall framework for the economic development and regeneration of a region as a whole" (House of Commons Select Committee on Environment, Transport and the Regions, 1998, p.ix).

The RDAs "will aim to spread the benefits of economic development across and within regions and through all social groups .... they will bring together programmes and expertise on physical and social regeneration in urban and rural areas and combine these with their wider work on training and business support and on enhancing the natural and built environment" (DETR, 1997, p.22). The White Paper that preceded the legislation setting up the RDAs explained that: "Rural needs and institutions may be different, but many of the same concerns — on skills, and access to training and to childcare, and on ways to foster new businesses — are common across each region. We need to understand the particular needs of rural areas, but to address them within an overall framework for the region as a whole" (p.24). It is therefore proposed that each RDA Board will include at least one member "who can contribute a strong rural perspective" (p.24). Finally, the White Paper specifies that "RDAs will design rural development programmes targeted on their most deprived rural area, and will monitor, consult and report on rural problems and how the agency is tackling them" (p.25).

As we have argued elsewhere (Lowe and Ward, 1998a,b), the understanding of rural issues in policy and plan-making in English regions has tended to be very simplistic and unnuanced. Rural policy has suffered from a lack of sensitivity to the 'differentiated countryside'. National agricultural policies have cast 'the countryside' as a unitary space. Current rural policy, indeed, tends to project onto a national stage a particular southern England model and agenda — largely to do with protecting the countryside from urban pressures. In consequence, uniform policies are applied that do not acknowledge regional differences.

The proposals for the new RDAs offer a window of opportunity to root rural development in regional economic realities, but only provided sufficient weight is given to rural issues. In this respect, rural policy has two main strands: the small-scale and the strategic. In the past, through such outlets as the Rural Development Commission and the last government's Rural White Paper (DoE and MAFF, 1995), the emphasis has been on the small-scale (i.e. village regeneration, community development and so on) but has been weak at the strategic (regional / national / EU) level. A key institutional challenge for the RDAs will be to remedy this weakness.

Integrating the rural in the regional

An inclusive regional policy would involve not only the geographical inclusion of rural spaces, but also a functional concern with the rural sector (sensitive to issues of scale/remoteness). A progressive agenda for reform calls for certain institutional and political developments at the national level as well as in the regions. Within the current structures of the DETR and the Ministry of Agriculture, Fisheries and Food (MAFF), changes will be necessary to reflect new priorities for agricultural and rural policy.

DETR, as the main territorial department within Government, will remain crucial in furthering progress towards sustainable development in the countryside. Its range of responsibilities puts it in a unique position to broker partnerships in order to develop and achieve cross-cutting policy objectives in areas such as land use planning, transport, housing and local government. Rural policy, however, does not tend to be one of its central preoccupations.

MAFF is the only ministry with a specifically rural focus and the loss of its food regulatory functions to the new Food Standards Agency does necessitate the renewal of its purposes and an overhaul of its organisation (House of Commons Agriculture Committee, 1997).

However, the producer-oriented structure of the CAP makes it difficult for MAFF to become a wholly different creature. Considerable efforts will have to be made alongside CAP reform to effect both cultural and organisational change within MAFF so that it has both the mission and competencies to act as an effective partner in furthering sustainable rural development. In preparation, MAFF will be required to re-examine its priorities and to establish a new departmental mission. This should involve a fundamental reappraisal and strengthening of MAFF's competencies on rural development and the rural environment. MAFF should be reformed and brought within the Government Office structures in the English regions. Such reform would provide a geographical/functional counterweight to the metropolitan orientation of other Government Office ministries, and help shift MAFF from a sectoral to a territorial ministry.

Co-ordination between MAFF and DETR at a strategic level could be assisted by the formation of the new Countryside Agency comprising the work of the current Countryside Commission and those functions of the Rural Development Commission not planned for transfer to the RDAs. This new body will need to establish a direct relationship with MAFF. It could monitor the rural performance of local and regional agencies, and guide the evolution of MAFF, and of the CAP into an Integrated Rural Policy.

A specific issue at the regional level concerns the detachment of MAFF's regional structure from that of the Government Regional Offices. The departments included are the Department of the Environment, Transport and the Regions, the Department of Trade and Industry, and the Department of Education and Employment. While it is obviously necessary for MAFF to retain its rural offices to service its client base, there does seem a persuasive case for incorporating the Ministry's regional hierarchy into the Government Offices. This would not only help overcome the urban bias in the Government Offices, but could also foster a more holistic and strategic approach to rural affairs within their regional contexts and encourage the reintegration of the agricultural sector into its local and regional contexts.

Conclusions

Debates about institutional reforms do not compete well for news coverage with stories of farm bankruptcies or collapsing livestock markets. However, the current co-incidence of CAP reform and the devolution agenda in the English regions provides an unprecedented and historic opportunity. The challenge for the North East is to look to the medium and long-term future of rural policy, with integration as the key organising principle.

At the time of writing, in December 1999, the new Regional Development Agencies have begun their work in the English regions. For a regional development strategy to be truly inclusive across a region's territory, sufficient attention will have to be given to the particular challenges posed by rural development questions. Theoretical debates around regional development have increasingly come to be characterised by a concern for endogenous development (from within), as opposed to exogenous development (from outside). The rationale for endogenous development is well-expressed in the Government's recent White Paper on competitiveness, which explains:

> In the global economy, capital is mobile, technology spreads quickly and goods can be made in low cost countries and shipped to developed markets. British business, therefore, has to compete by exploiting capabilities which competitors find hard to imitate. The UK's distinctive capabilities are not raw materials, land or cheap labour. They must be our knowledge, skills and creativity (Department of Trade and Industry, 1998).

Thus, in an age when so many factors of production are footloose, localities have to concentrate their efforts on where their strengths lie. In the North East, the region's rural areas are replete with resources, be they natural or cultural, and particular types of human resources. Like many other English regions, the North East's rural areas are one of the critical features that help define the distinctiveness of the region. In recognition of this argument, endogenous development strategies have for some time been pursued in the field of rural development. Examples would include the work of the Rural Community Councils, the Rural Challenge and Rural Development Programmes of the Rural Development Commission and in Objective 5b programmes under the Structural Funds. These are in contrast to some other development bodies in the North East who have stood accused of an over-reliance on attracting footloose inward investment.

In regional debates about endogenous versus exogenous development, the agricultural sector now has an exceptional and distinctive status. While agricultural policy is programmed at a European level, farming remains the one sector that is fundamentally and peculiarly dependent upon local conditions. It therefore seems absurd that the agricultural sector is not included within the remit of the RDAs. With the current round of CAP reform involving an increased role for national (and even regional) discretion in programming aspects of CAP expenditure, a constructive approach to agriculture in the English regions could assist in capturing and shaping this huge flow of annual resources currently going into the CAP. The challenge here would be to make CAP spending 'stickier' within the region.

A more inclusive and imaginative approach to regional territorial development that is sensitive to the rural dimension may also require a rethinking and redefinition of the concept of competitiveness, not least with respect to agriculture. Under a productivist CAP, competitiveness always tended to imply an individual farmer standing alone on world markets. Under such a definition, huge swathes of European agriculture would be uncompetitive without public subsidies. A more regional and territorial focus to agriculture could help to redefine agriculture's competitiveness within the context of a regional economy conceived of as a whole. As a result, agriculture can be attributed a wider role which acknowledges its multifunctionality. For as well as producing specific commodities, agriculture contributes to an attractive rural environment and the sustainable development of regions. Agriculture also serves as an input into other regionalised production chains such as tourism, locally distinctive foods, countryside recreation and other environmental goods and services.

Finally, vibrant and attractive rural areas can contribute to the overall quality of life in a region. In doing so, rural areas may help in embedding footloose industries and people in a region. There is recent evidence to suggest that skilled people and high growth firms are attracted to localities with attractive rural environments (Keeble and Tyler, 1995). All this is to suggest that an inclusive regional policy for the North East region, which is sensitive to the particular needs and attributes of rural areas, can yield social, economic and quality of life benefits for the region as a whole.

References

Commins, P. (ed) (1993) *Combatting Exclusion in Ireland, 1990-94: A Midway Report*, Brussels: European Commission.

Department of the Environment and Ministry of Agriculture, Fisheries and Food (1995) *Rural England: A Nation Committed to a Living Countryside*, London: HMSO.

Department of the Environment, Transport and the Regions (1997) *Building Partnerships for Prosperity: Sustainable Growth, Competitiveness and Employment in the English Regions*, Cm 3814, London: HMSO.

Department of the Environment, Transport and the Regions (1998) *Planning for the Communities of the Future*, Cm 3885, London: HMSO.

DTZ Pieda Consulting (1998) *The Economic Impact of BSE on the UK Economy*, Manchester: DTZ Pieda Consulting.

Dunn, J., Hodge, I. and Monk, S. (1998) *'Developing Indicators of Rural Disadvantage'*, paper presented at the Annual Conference of the Institute of British Geographers, Guildford, January 1998.

Errington, A., Harrison-Mayfield, L. and Jones, P. (1996) *Modelling the Employment Effects of Changing Agricultural Policy*, Reading Centre for Agricultural Strategy.

Hirsch, F. 1978. *The Social Limits to Growth*, Cambridge, Mass: Harvard University Press.

House of Commons Agriculture Committee (1997) *First Report*, HC Paper 310, Session 1997-98, London: HMSO.

House of Commons Select Committee on Environment, Transport and the Regions (1998) *Government Response to the First Report of the Committee: Regional Development Agencies*, HC Paper 645, London: HMSO.

Keeble, D. and Tyler, P. (1995) Enterprising behaviour and the urban-rural shift, *Urban Studies* 32, 975-997.

Lowe, P. and Ward, N. (1998a) *Regional Policy, CAP Reform and Rural Development in Britain: the Challenge for New Labour*, Centre for Rural Economy Working Paper Series No. 32, University of Newcastle upon Tyne.

Lowe, P. and Ward, N. (1998b) Memorandum of evidence, pp. 165-67 in House of Commons Agriculture Committee *CAP Reform: Agenda 2000*, HC Paper 311, London: HMSO.

Lowe, P., J. Murdoch, T. Marsden, R. Munton and A. Flynn 1993. Regulating the new rural spaces: issues arising from the uneven development of land. *Journal of Rural Studies* 9, 205-222.

Lowe, P., Clark, J., Seymour, S. and Ward, N. (1997) *Moralizing the Environment: Countryside Change, Farming and Pollution*, London: UCL Press.

Marsden, T., Murdoch, J., Lowe, P., Munton, R. and Flynn, A. 1993. *Constructing the countryside*. London: University College London Press.

Murdoch, J. and Marsden, T. (1994) *Reconstituting Rurality*, London: UCL Press.

Murdoch, J. and Ward, N. 1997. Governmentality and territoriality: The statistical manufacture of Britain's 'national farm', *Political Geography* 16, 307-24.

North East Regional Development Agency Contact Group (1998) *Regional Economic Strategy – A Preliminary Assessment*, Newcastle: Government Office for the North East.

Office of National Statistics (1998) *Regional Trends - 1998 edition*, London: Office of National Statistics.

Rural Development Commission (1993) *Rural Development Areas 1994*, London: Rural Development Commission.

Ward, N. and Lowe, P. (1999) *Agriculture and Rural Development in Northumberland*. Research Report. Morpeth: Northumberland County Council.

Woodward, R. (1996) 'Deprivation' and 'the rural': An investigation into contradictory discourses, *Journal of Rural Studies* 12, 55-67.

Woodward, R. (1998a) *Defended territory: the Otterburn Training Area and the 1997 Public Inquiry*. Centre for Rural Economy Research Report, University of Newcastle upon Tyne.

Woodward, R. (1998b) *Rural Development and the Restructuring of the Defence Estate*. Centre for Rural Economy Research Report, University of Newcastle upon Tyne.

10 Local Interests, Regional Needs or National Imperative? The Otterburn Question and the Military Training in Rural Areas

RACHEL WOODWARD

Introduction

The area of Northumberland encompassing Redesdale and the Upper Coquet valley, tucked up against the Cheviot Hills on the border between England and Scotland, is a place of many functions. Its designation as part of the Northumberland National Park indicates the value of this upland area in landscape terms. Environmental designations also denote the value of this space in terms of the habitats provided for rare flora and fauna. It has an agricultural function as an area dotted with hill farms where extensive agricultural practices, mostly livestock rearing, are practised. It is the destination for hill-walkers lured to the uplands by the promise of solitude in this vast and underpopulated landscape. It is also a space for military training. The Otterburn Training Area (OTA), provides the armed forces (mostly the army) with one of its best ranges for training soldiers in live firing and adventurous military training. It is a place of multiple uses.

This area is also a space of multiple meanings; it symbolises different things to different people. Walkers, farmers, naturalists and soldiers, not to mention the residents of villages along the rivers Rede and Coquet, all inscribe this space with meanings, values, attributes and significance. It is only natural that they should do so; the valorisation of place and space is a basic human activity. The study of those meanings and values is a requisite part of studies of conflicting land uses. Debates about land uses are factual arguments about what should go where, the functions a piece of land might serve, and the costs and benefits of competing functions. But they are also underpinned by more abstract arguments about the interpretation and meaning of place, as well as of the functions carried out upon it. In a place like Otterburn, which is used for many different purposes, the nature of competing land-uses cannot be studied without reference to the interpretations, meanings and arguments about significance imposed by different groups of people on that place and the activities carried out upon it.

Ordinarily, debates over the different meanings ascribed to a piece of land, be they in conflict or uncontentious, will not necessarily be visible beyond perhaps local discussions and comment in the local press. At Otterburn, debates over the use of this piece of upland and the meanings and interpretations underpinning those debates, have been made more visible. A non-statutory local public inquiry into developments proposed by the Ministry of Defence (MoD) for the OTA in 1997 brought into the open not only the range of arguments about the function of this place, but also the range of conceptual ideas underpinning those arguments about its meaning and significance.

It is the arguments over the interpretation, representation and meaning of the Otterburn Training Area that constitute the focus of this chapter. Of the many available arguments over the use and significance of Otterburn, I have chosen three to examine in detail here. After a brief introduction to the OTA and the 1997 public inquiry, I look first at the local significance of military land use, discussed by local residents in terms of the local economic impact of the OTA and the developments proposed upon it. Second, I look at the regional significance granted to the OTA area by conservation and amenity groups because of the natural and cultural heritage of this space. Third, I look at the national significance granted to this remote windswept corner of England by the Ministry of Defence. For each of these three arguments about the meaning and significance of the OTA, I look at both their form as they appeared during the inquiry, their possible origins among the actors promoting them, and question their wider consequences. The three sets of arguments, of the many made about the use of the Otterburn area during the public inquiry, have been chosen deliberately to reflect the ways in which the significance of the area's location in the North East was used in this particular land use debate. (For further information on the nature of the inquiry and the range of topics explored therein, see Woodward, 1998; 1999).

A brief introduction to the OTA and the 1997 Otterburn Public Inquiry

The Otterburn Training Area, covering some 22 per cent of the Northumberland National Park, consists of 22,908 hectares of upland moorland, rough pasture and wooded valleys. It is owned freehold by the Ministry of Defence, and as one of eight Army Field Training Centres is a vital area for the army for military training. Military activities on the OTA include live firing of artillery (there are two large impact areas into which live shells and rockets can be fired), small arms training, training in the use of pyrotechnics, dry training such as survival and orientation activities, low flying and parachute training. Some 30,000 soldiers (mostly British but also some from other NATO countries) train at Otterburn per year. The training area itself is largely uninhabited by civilians, save for 31 farms, the occupants of which are MoD tenants. Tenancy contracts encourage the pursuit of non-intensive agriculture, mostly grazing by some 28,000 sheep. Access around the training area is on single track metalled roads, some of which are open to the public. The training area also offers some of the best hill walking country in the north of England, although access by the public is restricted to specified rights of way and permissive footpaths, and is limited to around 150 days per year when live firing is not taking place.

In 1995, following two years of speculation, the army announced through a Notice of Proposed Development its intention to construct some additional infrastructure on the training area. The drawdown of troops from Germany under the Options for Change programme, a consequence of the changing geopolitical context for military activity in the aftermath of the post-1989 changes in Eastern Europe, meant the loss of training areas in Germany. Yet the army still required space to train using two of its heavy artillery weapons, the Artillery System 90 (AS90), a 45 tonne tracked self-propelled artillery system, and the Multiple Launch Rocket System (MLRS), a 25 tonne system for the launch of up to 12 rockets at a time, each with a range of up to 30km. Training in the use of these weapons systems requires an impact area for live firing. The army identified the OTA as the only one of its eight Army Field Training Centres as a suitable location. Some do not have a sufficient impact area for live firing. Others, Salisbury Plain, for example, were said to be full to capacity with existing military uses or, like Sennybridge in Wales, were deemed to have too small an impact area to ensure safety for personnel and civilians whilst live firing (see also Doxford and Savege, 1995).

However, despite the advantages of the OTA with its large impact area, the army faced a basic problem. First, the AS90 and MLRS are both very heavy and would sink into the soft peat of the upland area unless hardstandings (called gun spurs) were constructed to carry these machines whilst firing. These are stone pads of around 25m x 35m, connected to the roads of the training area by short metalled or stone tracks. Second, both weapons systems are deployed on a 'fire and manoeuvre' or 'shoot and scoot' basis; the enemy is targeted and fired upon, and then the weapons dispersed to avoid incoming retaliatory fire. In addition, these weapons operate at regimental level with a range of associated support vehicles carrying locational equipment, personnel and additional ammunition. These artillery systems and their associated support vehicles thus need to manoeuvre around the training area in order to function as intended. Not only would the soft peat of Otterburn inhibit this, but also the existing narrow metalled roads would not support the weight and width of such formidable weapons systems. Not unless roads were widened and additional tracks put into place could the weapons be used at Otterburn. The Notice of Proposed Development, its associated environmental impact assessments and a military justification for the choice of Otterburn, all published in 1995, proposed just that (MoD, 1995a - c). These documents set out the MoD's plans to construct some 24 new gun spurs and to modify 17 gun spurs, grouped into Gun Deployment Areas with associated Battery Echelon Areas for ammunition resupply, maintenance and other support functions through a circuit of track and a parking area concealed within woodland. Some 57.5km of existing roads and tracks would be widened to either 4m or 5m with an additional 1m hard shoulder, and an additional 15.3 km of new stone track would be built. Of this, 3.1km would form a network of linked stone tracks area joining up to 18 Tactical and 3 Technical Observation Post positions, stone lay-bys measuring about 40m x 10m. A 4.45 hectare Central Maintenance Facility would be built adjacent to Otterburn Camp and additional accommodation for 125 soldiers built within the camp itself.

After the publication of the MoD's plans, a considerable period of negotiation and consultation followed between the MoD and the National Park Authority, Northumberland County Council, various conservation groups and local residents. In the spring of 1996, the local planning authority, Northumberland County Council, rejected the plans put forward by the MoD and asked the then Secretary of State for the Environment to call the plans in for inspection within a local public inquiry. This request was granted, and from April to October 1997, the Otterburn Public Inquiry (hereafter OPI or 'the Inquiry') was held in Newcastle. The remit of the OPI was wide-ranging across an increasingly complex set of arguments about the costs and benefits of allowing the construction of infrastructure to enable a specific type of military training to take place at Otterburn. The MoD were the scheme proposers with support from some local residents and farmers. There were two principal groups of objectors to the scheme; the Northumberland National Park Authority working together with Northumberland County Council to object to the impact of this development on a National Park; and a Consortium co-ordinated by the Council for National Parks objecting primarily over the environmental impacts of the proposed developments, and including the Council for the Protection of Rural England, the Northumberland National History Society, the Northumberland and Newcastle Society, the Association of Countryside Voluntary Wardens, the Ramblers Association and the Open Spaces Society. In addition, a large number of individuals, locally and nationally, protested either in person or in writing about issues such as noise, pollution, traffic increases, environmental damage and damage to the peace and tranquility of this area.

The inquiry ranged far and wide across topics from the conservation of rare birds through to the vibration effects from vehicle convoys travelling to Otterburn. Issues of environmental protection, natural heritage, national security, public access, recreation, the use and abuse of the planning system and the attitude of the army to the countryside were all discussed. It is not my intention here to dwell on the range of debates at the inquiry. For the purposes of this chapter I am interested in one small element of the OPI, namely the arguments about land use which spoke also about the locational significance of Otterburn. It is to these that I now turn.

'Money in our pockets': the local economic significance of the OTA

The first set of arguments, chosen for what they say about the locational significance of Otterburn, are those produced mostly by local residents about the economic significance of the Training Area. In a way, the economic arguments about the developments proposed for the OTA were a non-issue. For example, during the pre-Inquiry period, I was approached by both a local landowner and a member of the County Council to discuss how the economic impact of the developments could be debated at the OPI. These arguments were soon dropped when it became apparent that the economic impact was not going to be a major issue in the debate. The OTA currently employs about 120 people, mostly civilians and including seasonal and contract workers, for a variety of range management tasks. The OTA is therefore an important local employer. Were the developments to proceed, about 14 additional jobs would be created. Were there to be no development, business would continue as usual with the OTA remaining a significant local employer. There was never any suggestion that the OTA would cease to function as a major army training area if the developments did not go ahead as planned, and once this fact became apparent at the early stages of the Inquiry, no group or individual tried to sustain an argument on employment grounds either for or against the development. It was, quite simply, not a big issue.

However, the economic significance of Otterburn stayed on the agenda as an issue for debate, lurking in the background and appearing from time to time when the impact of the development in purely local terms was discussed. After all, around 30,000 soldiers a year come up to Otterburn to train. Local cafes, restaurants, pubs and shops all benefit substantially from the patronage of passing armies. The OTA is good for the businesses of Otterburn village. For this reason, many Otterburn residents were reluctant to oppose the developments. Although the MoD consistently denied that the proposed developments would lead to increased use of the ranges, in the minds of many local business people the proposed developments logically meant extra business.

It is difficult to quantify levels of local support or opposition to the proposed developments at Otterburn. Those in favour of the developments claimed near-total support at a local level from their fellow-residents, a level apparently gauged through signatures on a petition presented to Parliament in February 1995 with the assistance of the local MP. Those against the developments complained of intimidation, real or threatened, in silencing the local voices of opposition. I would argue that the economic benefits of the OTA played an important part in mobilising local support for the developments. The local MP certainly thought so. Writing to the National Park Authority in September 1995, he stated his belief that "the military's plans are essential in the national interest and also for the continuing economic prosperity of people living in the locality of the ranges". The petition to Parliament sought to remove National Park status from the OTA, on the grounds that this designation was not in the best interests of either national defence or the local economy. The petition went on to argue that the National Park Authority's decision to object to the army's proposed developments

> [...] poses a very serious threat to the economy of the area which benefits by over 6 million [pounds] annually from the Army's presence, through direct civilian employment, letting of farms, purchase of services and spending by military (Hansard, 1995).

This was certainly the view of one local District Councillor in a statement to the OPI:

> It is estimated that 90 per cent of the local people support the Army presence, recognising the contribution it makes to the local economy through employment, directly or indirectly. The £700,000 spent annually on the maintenance of its rural estate is contrasted to the £189,000 spent by the National Park on conservation measures. It must be acknowledged that the additional income and employment opportunities offered by the present proposals will be very welcome to the local economy. Local people, therefore, view the proposals against this background (Alnwick District Councillor for Harbottle Ward, OPI, 3rd June 1997).

This statement contains some factual sleight-of-hand, of course. The figure quoted for maintenance is for the rural estate across Britain, to be used for a variety of range maintenance projects. It is not a useful figure for comparison with the amount of money spent by the National Park Authority. Its use here served a dual political function of quiet criticism of the National Park Authority while drawing attention to the financial input of the army in estate management. The key points that this quotation is trying to establish are that the local economic benefits of the OTA should not be overlooked, and that the OTA was primarily significant for local residents not because of environmental attributes or its value as a protected landscape, but for its economic significance to the area.

In pointing this out, I am not implying criticism of this particular strategy for portraying the significance of the OTA. Indeed, this strategy is perfectly understandable. First, this is an area with a fragile rural economy under pressure from declining agricultural opportunities and increasingly dependent on the vagaries of the tourist trade. Of course, people will want to encourage whatever brings finance into the area. Second, I suspect that many residents of Otterburn village looked down the A696 to post-industrial Tyneside and spoke of their fear of the destruction of their own economic base (the army ranges) by drawing parallels with the effects of the closure of the mines and shipyards of the industrial North East. There were good reasons for promoting the significance of the OTA locally in economic terms.

But there was dissent, of course. Some argued that the proposed developments posed a very real threat, couching that opposition partly in economic terms too. For example, one resident of Otterburn village spoke to the Inquiry of the dangers posed to the local economy by the army's dominance of it. An effect of the proposed developments, she argued, would be to further local dependence on the military, and to consolidate an existing employment pattern which favoured male civilian employees of the MoD because of the gendered nature of much of the work on offer. In addition, she argued that the threats posed by the proposed developments to the tourist trade would reduce still further the employment prospects for women in the area (OPI, 22nd July 1998). A resident of Harbottle on the other side of the training area developed further this idea that damage to the tourist industry could follow the developments:

> People are put off coming here because they are not sure where they are safe to walk or drive. With night or weekend firing on the increase they are unlikely to be attracted to stay very long. Tanks by the roadside are not a welcoming sight. The Army do not appear to be offering much in terms of extra civilian employment so for many of us tourism is the only way we

have of making a living. But if the tourists don't come because of the loss of tranquillity and access, we residents suffer financially and our children will have no option but to leave (OPI, 23rd July 1997).

Another resident, while recognising the importance of job security promised by the developments for those economically dependent on the army, emphasised that "there are many other constructive ways of employing people in the countryside" beyond military employment (Letter to OPI, 7th October 1997).

The language in which support or dissent was voiced for the proposed developments and their economic significance was also indicative. Discourses of both rurality and localism were drawn upon on all sides. For example, the District Councillor quoted above in support of the military input into the economy made a clear distinction between those who in her mind were of the countryside, and those who were not:

> [...] I must challenge a perception which is advanced and espoused by those who wish to be associated with the countryside, but who do not earn their living from the land. That is that visitor access and conservation are compatible, but that military activity and conservation are not (Alnwick District Councillor, Harbottle Ward, OPI, 3rd June 1997).

The use of a specific discourse of rurality — the linguistic framework in which ideas about the meaning of rural areas are constructed — echoes that drawn upon in the local petition presented to Parliament in 1995. This also constructed an idea of what it is to be rural with reference to the priorities which should be accorded to local knowledge:

> The Park Authority, with a significant proportion of its members coming from urban areas, has demonstrated little understanding of the way military activities, compatible with hill farming, have become part of the rural way of life in the Otterburn area for the past 80 years (Hansard, 1995).

This discourse of rural specificity, with its emphasis on self-determination and the power of local knowledge appears time and again in the letters to the Inspector from residents of the area, arguing that there was some intrinsic appreciation within the MoD of rural needs, which bodies such as the National Park Authority somehow failed to recognise. This quotation is an example of a common lament:

> We can't do without the MoD who look after our interests better than the National Parks because it understands the needs of the local people and our way of life (Letter to National Park Authority, February 1997).

Elsewhere, discourses of localism were drawn upon to shape arguments that local residents themselves should determine the fate of the training area, and that outsiders (defined in a variety of ways) had no grasp of the key issues at stake. In the words of the Licensee of a pub in Upper Coquetdale:

> I object strongly to ... private individuals who do not live in this area who wish to dictate as to what is good or bad for the National Park (Letter to OPI, 19th February 1997).

Some even argued that only those born and bred in the area had the authority to speak on the impacts of the proposed developments. The implication of many of these statements was that the fact of residence or length of habitation in an area determined those who had a 'right' to speak with regard to the developments and those who had no such authority. Furthermore, long-term residence was constructed as an automatic endorsement of the activities of the army on the Otterburn ranges. For example:

> There are no newcomers to the area who were alive when the Range started — so all were forewarned. There are few descendants of pre Range occupiers and even fewer of them are objectors. Everyone living in the area is aware of the range and its use (Statement to OPI, 3rd June 1997).

These tactics, of establishing rural residents as the sole source of authority on issues of rural land use, and of establishing the idea of 'locals' as the only group with a natural right to speak on local development issues, were hard to counter. One objector to the proposed developments spoke directly on this point and its possible consequences for public debate:

> I fear I will be one of very few local people who feel free to speak because I am not beholden to the MoD for my employment or my tenancy. I do risk being ostracised by neighbours who by right of birth or longer term of residency consider me yet another outsider who has not right to an opinion on activities in this area or the freedom to voice it. However, whether my family has been here for generations or whether I arrived yesterday is of no consequence. What is important is that I am free to speak and who knows, I may be speaking up for those who are constrained from speaking (Statement to OPI, 23rd July 1997).

In conclusion, then, it seems that a dominant line of argument appeared in the debate over the Otterburn which talked of the army as a significant and positive force for economic development in the Otterburn area, and furthermore, that the inherent understanding of rural economic development issues of people who lived and worked in the immediate area consolidated these claims. This was a very dominant argument — there were even claims (unproven at the time) that the MoD had been instrumental in establishing a local group, the self-styled 'Association of Rural Communities' with the specific intent of promoting this localist, ruralist argument about the economic significance of the military. Whatever the truth of the matter, it seems clear that the locational significance of Otterburn was established at the local level in economic terms.

'The Land of the Far Horizon': Regional distinction and specificity

In this section I turn to the regional significance of the Otterburn Training Area, and to a different group of participants in the public debate over the proposed developments. The facts of Otterburn's specific natural and cultural heritage, and of its location in the North-East of England, were very important to opponents of the scheme. This regional distinctiveness and specificity was established in order to argue that the army's plans would destroy the special and unique contribution made by this place to the cultural landscape of the North East.

The natural heritage of the OTA is undoubtedly precious. No party to the debate over Otterburn's future ever tried to deny this. The upland moorland landscape is rich in rare fauna and flora. Eleven Sites of Special Scientific Interest (SSSIs) have been designated on the OTA in recognition of this. The area lies within the Northumberland National Park, designated in 1956 in recognition of its outstanding landscape qualities. For many in Northumberland, the Park and particularly its northern reaches away from the honeypot of Hadrian's Wall provides a playground for the region, a place for recreation, exercise, enjoyment and spiritual refreshment away from the noise and bustle of urban life. Furthermore, the relative remoteness of this part of the National Park, and the relatively small numbers of visitors to it were facts drawn upon to emphasise the qualities of peace and tranquillity and the protection this afforded to rare birds and insects. For example, the Large Heath Butterfly and the Black Grouse both flourish in the boggy uplands despite (or, for the army, because of) the deterrent effect of live firing scaring off human predators and their dogs. The regional significance of this place was emphasised in terms which celebrated this natural heritage. For one Newcastle resident, writing to the Inspector towards the end of the Inquiry, this was sufficient to warrant objections to the army's proposed developments:

I hope it is not too late for me to express my concern about the military proposals for the Northumberland National Park. The reason that living in the city is bearable, (the city of Newcastle), is that there is Northumberland to escape to. Even if one is unable to go there, it makes life bearable to know that there *is* fresh air, there are rare bogs, there are lovely wild birds, there are beautiful clear rivers, in the National Park (Letter to OPI, 2nd October 1997).

The cultural heritage of the OTA was also used to emphasise the regional specificity of Otterburn and to argue for its valorisation and preservation. One strategy used by members of the Council for National Parks Consortium was to stress the area's unique history. Its troubled past as a border outpost and marginal place for the Roman Empire, and as a border zone between the warring nations of English and Scots, was used to construct the idea of Otterburn as a cultural landscape unique in Britain. In the words of one objector to the scheme:

The historical associations are those of a wild Border landscape, the setting for Border raiders and (with the Scottish side of the Border) of the best-known of British narrative ballads. Redesdale, on the English side, was a heartland of the clans of Border farmer-thieves. It is likely that the influence of Border insecurity in inhibiting economic development for several centuries contributed to the wilderness qualities of the Northumberland National Park, and makes the moorland and semi-natural woodland landscape of the OTA relict and therefore of historic importance in two senses. It is relict from the 16th and 17th centuries through to the 20th century (the bastles are a part of this), and it is relict within the 20th century while surrounding moorlands were afforested or experienced agricultural intensification (NPC/P/1a, p.7).

A further strategy in the construction of this historical narrative was to emphasise the sheer variety of cultural activities which had left their inscription on this place. Whereas the MoD's Proofs of Evidence to the inquiry told the history of Otterburn as the history of military use of the area, those of the CNP Consortium promoted the idea of a much longer chronicle of human use, with people in each phase leaving their inscription on the landscape. The history of Otterburn became one of the interplay of human and environmental events, rather than one of 20th century human impact. The landscape therefore could be valued less for the linear history written across it than for the variety of imprints it bore. In establishing the history of the OTA in this way, it was then possible to argue for the destructive potential of the proposed developments. If there were nothing there, the developments would cause no harm. But a landscape with a rich cultural heritage could easily be destroyed by the construction of military infrastructure.

Regional distinctiveness and specificity was also to be found in the atmosphere and feel of the Otterburn area. Much of this was attributed to the physical qualities of the landscape which were presented as provoking a specific spiritual response. Use was made of the writings of the historian G.M. Trevelyan who "catches the spirit of the place in splendid prose":

> In Northumberland alone, both heaven and earth are seen; we walk all day on long ridges, high enough to give far views of moor and valley, and the sense of solitude above the world below, yet so far distant from each other, and of such equal height, that we can watch the low skirting clouds and they 'post o'er land and ocean without rest'. It is the land of the far horizons. (G.M. Trevelyan, *The Middle Marches*, 1938, quoted in NPC/P/3, p.8).

Indeed, Trevelyan was used in a number of different contexts during the course of the Otterburn debate. In this instance, the emphasis was placed on the ways in which this landscape had 'long inspired a sensitive human response" (NPC/P/3, p.9). The spiritual response to the landscape gave the area its value. Trevelyan was quoted in support of this argument too:

> ...it is no less essential [...] to preserve for the nation walking grounds and regions where young and old can enjoy the sight of unspoiled nature. And it is not a question of physical exercise only, it is also a question of spiritual exercise and enjoyment. It is a question of spiritual values. Without vision the people perish and without sight of the beauty of nature the spiritual power of the British people will be atrophied (Trevelyan to the Standing Committee on National Parks, 1938, quoted in NPC/P/1, p.9).

Through the use of quotations such as these, a link could be made between the landscape qualities of this part of Northumberland and a national need for open space. Otterburn was presented as unique in respect of the qualities of the landscape which enabled people to do this:

> The Otterburn ranges are one of very, very few places in England where one can walk for hours without seeing any recent works of man. Even the more beautiful hills of the Lake District cannot offer this to nearly the same degree. It is not an untouched wilderness; there is no such place in these islands. That concept depends on the walker's awareness of the history of our countryside or at least on his desire to escape from our pervadingly man-made surroundings (Letter to OPI, 30th September 1997).

The developments proposed by the army for the Otterburn Training Area would destroy this wide open area of wilderness:

> For the construction period this part of the Park [i.e. the OTA] would be effectively unusable by visitors. When complete the visitors would be unable to move anywhere by car without being confronted by a new scene. There would be major impacts at the tactical O.P.s, the Otterburn Camp and the Airfield, all of which would be highly visible in the local landscape and the atmosphere, the spirit of the place, would be dominated by the evidence of mans' intervention. Whereas at present there is still an opportunity to experience the feeling of wilderness over much of the area, that would have been destroyed. The balance, in short, would have been tipped irrevocably and the essential qualities of this part of the Park would have been lost forever (NPC/P/3, p.16).

Otterburn, then, had particular significance because of its cultural and natural heritage and the landscape that this had produced. These landscape attributes were drawn upon by opponents of the proposed developments. They argued that this space had regional significance because of what it offered to the residents of the region in terms of open countryside, facilities for walking and exercise, and above all, opportunities for spiritual refreshment. Furthermore, its cultural history made it unique as a border area, contributing to the specificity of the North East. Any attempts to destroy this specificity, as opponents of the proposed developments argued the developments would, would remove from the region one aspect of its distinctiveness.

National imperative: front line or remote corner?

The third set of arguments I want to discuss, chosen for their locational significance, were those produced by the army (in fact, by the MoD on behalf of the army) about the importance of Otterburn at a national level. In the course of this set of arguments, the OTA shifted in the portrayals presented by the MoD, from being a far-flung training ground in a remote corner of northern England, to being the nexus for military training which in turn meant that Britain's defensive capabilities were ensured. Otterburn was portrayed as locationally significant in serving the national interest in strong defence. Furthermore, this national defence interest was promoted as the most important land use function served by the training area. Although conservation and the protection of the landscape qualities of this National Park were deemed significant in the evidence of the MoD to the Inquiry, they were of less importance than national defence. One could argue, of course, that this is hardly surprising. The institution with the obligation to defend the realm would surely not argue otherwise? This is true. But my argument is not that we should find the MoD's emphasis on defence surprising, but that we should look carefully at the way that defence role is portrayed for what it says about the locational significance of the OTA.

The MoD has consistently portrayed the OTA as a place for army training, significant primarily for the soldiers who happen to be there. "The MoD holds its estate at Otterburn for one reason — to provide realistic training facilities for the Armed Forces" says a conservation management plan (MoD, 1990, p.8). A similar document published three years later expanded on this theme:

> The military training requirement is the paramount concern in the management of OTA and it must be recognised that if the training area is to provide the facilities to train our modern Army to an acceptable standard, it cannot be frozen in time, but must be allowed to develop to accommodate the changing requirements cannot (MoD, 1993, p.84).

In these and other documents, the portrayal of Otterburn is as a military training area, nothing more and nothing less. However, when the use of that space for military training and the consequences started to be questioned, most notably at the public inquiry, the ways in which the MoD presented the locational significance of the OTA shifted. It started to argue for the OTA as a central location for British defence policy, without which the UK would be unable to fulfil its overseas and domestic defence commitments. Otterburn shifts from being a remote windswept rural training area to being the front line in UK defence policy.

One catalyst for this shift was the priority accorded to Otterburn under changing military roles in the 1990s. The first MoD Proof of Evidence presented to the OPI opened with the following assertion:

> The Army must meet the challenges of today and tomorrow, not yesterday. The transformation in the international security environment following the end of the Cold War has led to a major reassessment of the UK's defence strategy and policy, and the Army has adapted its command and force structure accordingly. A much higher premium is now placed on the deployability and flexibility of our forces (MoD/P/1, p.4).

In arguing this case, much was made of the variety of commitments under the Joint Rapid Deployment Force (JRDF) and to NATO under the Allied Command Europe Rapid Reaction Corps (ARRC) within this post-Cold War context. The AS90 and MLRS regiments were identified as part of these commitments, to be maintained at high levels of readiness should the need for their use arise. "Precise details of the readiness levels of particular units are classified", said one of the MoD's spokespeople, but it was argued that these regiments needed to be ready to be deployed in a matter of days. "The training requirement arising from the need to deploy at very short notice is considerable." The argument went on:

The consequences of the Army proposals for development of Otterburn not being implemented are therefore as follows:

It would delay the time when forces assigned to NATO or the JRDF were declared ready to conduct operations, whilst necessary pre-deployment training was conducted, or it would require units to conduct in theatre training (if facilities exist). It would reduce the operational effectiveness of the force which could lead to jeopardising the successful outcome of the mission. The standard of training of the Royal Artillery would be degraded with a consequent definite and adverse impact on capability and morale (MoD/P/1/S, pp. 11-12).

Another catalyst for this shift in the significance of Otterburn was the terms of the inquiry and its emphasis on the interpretation of planning guidance, specifically Planning Policy Guidance Note 7 (PPG7) on land use planning in England's rural areas. Paragraph 4.5 of PPG7 states that major development should not take place in National Parks, save in exceptional circumstances where they can be demonstrated to be in the public interest before being allowed to proceed. Accordingly, the proposed developments to Otterburn had to be demonstrably in the public interest, gauged in terms of the needs and alternatives to the scheme in question. It followed that the MoD had to demonstrate the national need for their proposed developments. One tactic chosen to demonstrate this need was the emphasis placed on the priority which should be accorded to national defence. Use was made, for example, of the House of Commons Defence Committee's (HCDC) statement that:

Failure to allow the range to be used by these weapons would severely limit the ability of the Royal Artillery to obtain the practice firing which it needs to retain its operational capability. [...] it is in the National Interest for the range of improvements at Otterburn to be allowed to proceed (HDCD, 1996).

The national interest is therefore defence. Other concerns, such as conservation of the natural environment were important but secondary to this:

The great importance which the Government places upon environmental interests should not obscure this fundamental importance of the proposals (I/MoD/292, p.6).

This national importance, the need for military training, was then constructed as unassailable, perhaps unarguable:

> We do not expect anyone to deny this nor do we expect it to become an issue, because failure to be properly trained in the use of these weapons means greater difficulties in succeeding in military conflict objectives and more causalities. The proposals now being put forward [...] are to enable those who use, and will use, these weapons to be fully and properly trained with the weapons and to be at required readiness states. So we say that the development proposals are of the highest national importance and self-evidently in the public interest. I say that this is beyond argument (I/MoD/1, p.2).

This last quotation from the opening statement by the MoD at the start of the Inquiry effectively set the tone for the debate, in which discussion of national defence interests was notable by its absence. The multiple uses and meanings of Otterburn that were debated and disputed at the Inquiry always sat within the context of a broader discourse which granted national significance to Otterburn, and prioritised this over other concerns.

This repositioning of Otterburn as the first line of defence, and the positioning of defence as the most important of a range of national interests went mostly unchallenged during the public inquiry. Personal military connections, oppositional political strategies, public inquiry tactics and individual choices meant that few participants in the inquiry challenged the MoD on the broader question of *why* the armed forces do what they do on the Otterburn ranges in the first place. An elderly Tyneside resident wrote to the National Park Authority in 1995, at the outset of the public debate over the future of Otterburn. Her letter is unusual in bringing to the fore what others ignored:

> Father left the best years of his life in the Western Front, he was 31 years old when he got back home. His two sons my eldest brothers were killed in bomber command. Surely they didn't die to create death for some-one else and to deprive our people of some lovely landscape (Letter to National Park Authority, June 1995).

Conclusions

In this chapter I have chosen three sets of arguments and tried to show how each spoke about the locational significance of the Otterburn Training Area. I have argued, first, that at a local level the locational significance of the OTA was portrayed mostly in economic terms. These were not the only terms in which the place was discussed; others included the peace and quiet of living in a rural area, the possible disruption caused by heavy artillery firing, and threats to the high quality of the natural environment from pollution. But framing the debate about the future of the OTA in economic terms meant that connections could be drawn between the Otterburn locality and the rest of the North East. The experience of economic devastation on post-industrial Tyneside could then illustrate the consequences of the loss of the economic base. Second, I examined how the cultural heritage of the area was portrayed as distinctive, and how this specificity used to counter the claims of the site developments. The proposed developments, rather than having a negligible or minor impact on the cultural and natural heritage of the area, were portrayed as potentially destructive of the things that were most unique and special about the Otterburn area, and by extension, the North East. Third, I argued that Otterburn was promoted by the army and MoD in terms which emphasised its national significance as a key nexus in much wider structures of defence organisation. Otterburn was not only crucial to the British army in order to provide training but also critical to NATO in terms of the contribution it made to wider military missions. Again, by extension, the North East was constructed as symbolically important within wider national and international defence strategies.

 The final point to make concerns the future. The meanings that we ascribe to different places are fluid and unstable. For example, I have tried to show how the MoD's portrayal of the OTA shifted somewhat over the 1990s as part of a wider strategy to emphasise the centrality of the OTA in defence planning. They change with the requirements of those promoting them. The point about change is an important one to make in this context, because although the OPI closed in October 1997, in 1999 the Department of Environment, Transport and the Regions decided that further information revealed during the 1998 Strategic Defence Review made it necessary to re-open the Inquiry. If the proposed developments proceed, it is also likely that again the meanings of the Otterburn and its locational significance will shift to take account of the subtly different uses made of the training area. We shall see. The meanings we give to places are never immutable.

References

Doxford D. and Savege J. (1995) The proposed development of the Otterburn military training area in Northumberland National Park: a national perspective. *Journal of Environmental Planning and Management* 38, 551-560.

Hansard (1995) Petitions: Otterburn Military Training Area: Mr Peter Atkinson (Hexham). *House of Commons Hansard* Vol. 254, Col. 565. 10th February 1995.

House of Commons Defence Committee (1996) *Seventh Report: Statement on the Defence Estimates.* Session 1995-96. HC 215. HMSO, London.

I/MoD/292 (1997) *Closing Submissions on Behalf of the Ministry of Defence.* OPI Document.

Ministry of Defence (1990) *OTA Conservation Management Plan.* Unpublished OTA document.

Ministry of Defence (1993) *OTA: Strategic Estate Management Plan.* Unpublished OTA document.

Ministry of Defence (1995a) *Otterburn Training Area: Options for Change Proposals. Explanation of the Proposals and Non-technical Summary of the Environmental Statement.* Ministry of Defence and RPS Clouston, Abingdon.

Ministry of Defence (1995b) *Otterburn Training Area: Options for Change Proposals. Environmental Statement.* Ministry of Defence and RPS Clouston, Abingdon.

Ministry of Defence (1995c) *The Military Justification for the Development of Otterburn Training Area to Accommodate AS90 and MLRS Training.* No further publication details available.

MoD/P/1/S (1997) *Proof of Evidence on UK Army Training Requirements, presented by Lt. Col. James Carter for the MoD.* Presented 22nd April 1997. OPI document.

MoD/P/1/S (1997) *Proof of Evidence on UK Army Training Requirements, presented by Lt. Col. James Carter for the MoD.* Summary. Presented 22nd April 1997. OPI document.

NPC/P/1 (1997) *Proof of Evidence for the Council for National Parks, presented by Vicki Elcoate.* OPI Document.

NPC/P/1a (1997) *Proof of Evidence for the Council for National Parks, presented by Dr Angus Lunn.* Presented 17th September 1997. OPI Document.

NPC/P/3 (1997) *Proof of Evidence for the Northumberland and Newcastle Society, presented by Graham Coggins.* Presented 17th September 1997. OPI Document.

Woodward R. (1998) *Defended Territory: The Otterburn Training Area and the 1997 Public Inquiry.* Research Report, Centre for Rural Economy, University of Newcastle upon Tyne.

Woodward R. (1999) Gunning for rural England: the politics of the promotion of military land use in the Northumberland National Park. *Journal of Rural Studies* 15, 17-33.

Printed and bound by CPI Group (UK) Ltd, Croydon, CR0 4YY

21/10/2024

01777082-0002